For Karen and for Ruth

Contents

List of Figures

List of Tables

Acknowledgements

Our first thanks go to our colleagues in the Computer-Aided Design research group at UC-Berkeley, known locally as cadgroup, although it is by no means the only group working on CAD at Berkeley. Over the past few years, cadgroup has been home each year to over 60 graduate students and industrial visitors, working on everything from behavioural synthesis to compaction to device simulation to analog CAD, and a great deal in between. Special thanks go to the other cadgroup faculty: Alberto Sangiovanni-Vincentelli, Richard Newton, and Don Pederson, who founded cadgroup and continue to make it one of the world's most exciting and pleasant places to work. We'd like to say a special thanks to the logic synthesis group, and in particular to Rick Rudell, Albert Wang, Sharad Malik, Alex Saldanha, KJ Singh, Tiziano Villa, Ellen Sentovich, Antony Ng, Hamid Savoj, Cho Moon, Luciano Lavagno, Herve Touati, Yosi Watanabe, Paul Stephan, Tim Kam, Paul Gutwin, Will Lam, Narendra Shenoy, Rajeev Murgai, Masahiro Fukui, and Abdul Malik.

Cadgroup software is largely supported by students; almost all the software we use is kept going by one or more of us – this includes the document processing software used to write this book. This is a great deal of effort greatly appreciated, so thanks to Rick Spickelmeier, Tom Quarles, Tom Laidig, Dave Harrison, Rick Rudell, Beorn Johnson, Brian Lee, Chuck Kring, Gregg Whitcomb, Wendell Baker Luciano Lavagno, Andrea Casotto and Don Webber for generously pitching in.

Cadgroup isn't entirely student-supported. Thanks to Brad Krebs and his staff, Mike Kiernan, Valerie Walker, and Kurt Pires, who keep an installation of three mainframes and over fifty workstations up and running. The never-ending stream of paperwork and budget balancing keeps our clerical and administrative staff busy. Our thanks to Shelly Sprandel, Flora Oviedo, Irena Stanczyk-Ng, Maria Delgado-Braun, Erika Buky, Elise Mills, Susie Reynolds, Sherry Parrish and Deirdre McAuliffe-Bauer.

We would like to acknowledge the support of the staff at the Computer Science Department of the University of British Columbia, who maintained the software at that end of the publication and kept the communication lines open: Carlin Chao, Koon Ming Lau, Marc Majka, Peter and George Phillips, Grace Wolkosky, Evelyn Fong and Gale Arndt.

We'd also like to thank our main sponsors through the course of this research: the Semiconductor Research Corporation under contract DC-87-008 who supported most of this research, and to the Digital Equipment Corporation for generous equipment grants over the years. For the preparation of the manuscript, we thank the Natural Sciences and Engineering Research Council of Canada and the University of British Columbia.

Rick McGeer would like to thank his parents, Pat and Edie McGeer, for their years of support, encouragement, and advice.

This book is dedicated to Karen and Ruth, without whose endless support this work could not have been done.

Preface

This book is an extension of one author's doctoral thesis on the false path problem. The work was begun with the idea of systematizing the various solutions to the false path problem that had been proposed in the literature, with a view to determining the computational expense of each versus the gain in accuracy. However, it became clear that some of the proposed approaches in the literature were wrong in that they underestimated the critical delay of some circuits under reasonable conditions. Further, some other approaches were vague and so of questionable accuracy. The focus of the research therefore shifted to establishing a theory (the viability theory) and algorithms which could be guaranteed correct, and then using this theory to justify (or not) existing approaches. Our quest was successful enough to justify presenting the full details in a book.

After it was discovered that some existing approaches were wrong, it became apparent that the root of the difficulties lay in the attempts to balance computational efficiency and accuracy by separating the temporal and logical (or functional) behaviour of combinational circuits. This separation is the fruit of several unstated assumptions; first, that one can ignore the logical relationships of wires in a network when considering timing behaviour, and, second, that one can ignore timing considerations when attempting to discover the values of wires in a circuit.

The failure of the first assumption is manifested by the false path problem. The failure of the second is manifested by the failure of naive solutions to the false path problem, which ignore the element of time in determining the logical behaviour of the wires in the circuit. The consequence of this is that naive attempts to describe the logical or temporal behaviour of integrated circuits fail, in general, to describe either. The lesson to be learned is that the two concepts are inseparable and must be described together. This conundrum forms the essential mystery of the time domain in the analysis and design of integrated circuits. It is our hope that this book, which describes the full solution to the false path problem, can shed some light on the general problem of the time domain in integrated circuits.

What follows here is a very brief, informal treatment of the contribution of this book.

The False Path Problem

One major problem in the design and analysis of integrated circuits is the following: how long does a circuit take to compute its function? There are a variety of ways to answer this. The most popular has been to model the circuit as a network of capacitors and variable resistors and then run a circuit simulator such as SPICE; however, the problem with such simulators is to determine the sequence of input waveforms that manifest the longest delay in the circuit. Further, a SPICE run involves solving a system of differential equations of size equal to the number of capacitive nodes in a circuit. Running SPICE is obviously impractical when the number of nodes in the circuit approaches the number commonly found in modern VLSI circuits.

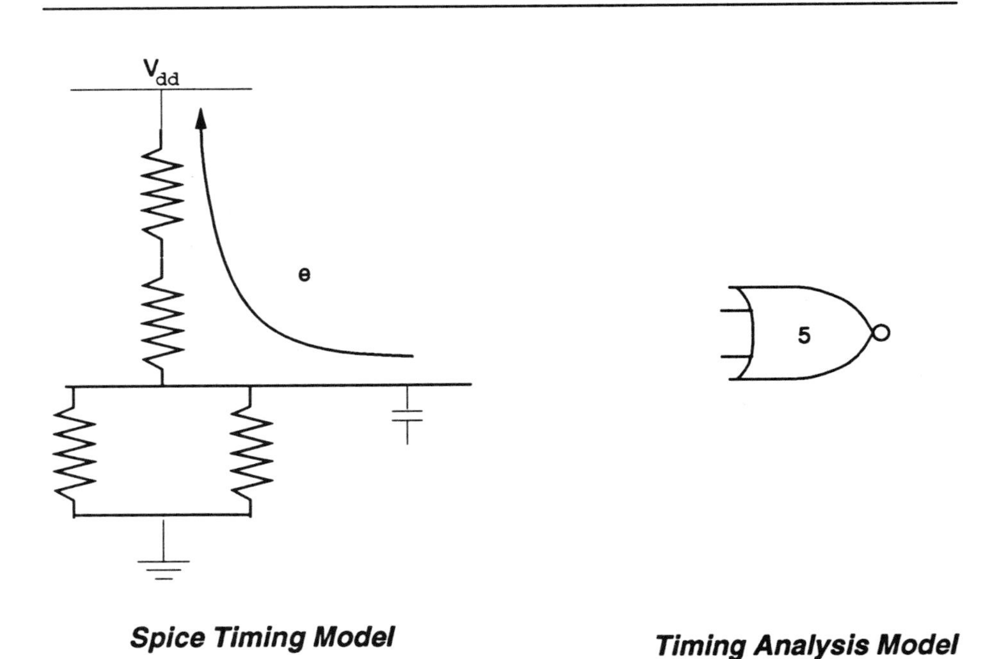

Figure 0.1: Different Timing Models

Therefore, as VLSI circuits grew in size, new methods for answering this question came into vogue. In particular, it became popular to model the circuit not as a collection of resistors and capacitors, but, as a graph of logic gates, where each gate has an associated with delay. The modeling or translation between the different levels of abstraction is illustrated in figure 0.1. Under this more abstracted model, the delay from an input to an output is the delay of the longest (in the delay sense) connected sequence of gates, or *directed path*, between the input and output. The problem of finding the delay of the circuit is then simply the problem of finding the longest path through a weighted, directed acyclic graph, a very simple problem. The difficulty, however, is that many paths turned out to be false.

False paths arise because the time that we wish to delay the clock is not governed by the length of a path through a graph, but, rather, by the length of time between the time the inputs to the circuit arrive and the outputs to the circuit settle to their stable values. This latter time is obviously the time of the last event on the outputs. All events originate at the inputs, and travel from the inputs to the outputs via paths through the graph. Hence we set the clock to the length of the longest path to allow for an event to travel down this path. However, not all paths can propagate events, and we should only set the clock to the length of the longest path that may propagate an event. Paths that cannot propagate events are called *false*, and identifying the longest path that can propagate an event is called the false path problem.

An example of a false path is shown in figure 0.2, which can also be found in chapter 1. For an event (a change in value) on x to propagate to a, we must have $y = 1$. For a to propagate to b, we must have $z = 1$. But for b to propagate to c, we must have $y = z = 0$. Hence the path $\{x, a, b, c, d\}$ *appears* to be *false*.

Appears, however, is the operative word here. When we say we must have $y = 1$ for x to propagate to a, we haven't specified one key detail: we haven't said *when* we want $y = 1$. Obviously, we only need $y = 1$ when we are attempting to propagate x to a, that is, at $t = 0$. Similarly, we're trying to propagate b to c at $t = 2$ (assuming unit delay on all gates), and so we only need $y = z = 0$ at $t = 2$. Hence this path is *true* if we assume y switches from 1 to 0 at $t = 0$ and z switches from 1 to 0 at $t = 1$. However, if we know that y and z both settle at $t = 0$, then the path is false.

The first innovation of the viability theory, presented in this book, is to take into account the changing nature of the signals in the circuit; that is, to permit a signal y to be at different values at different times t_1 and t_2 if there was a possibility of a switching event on y between t_1 and

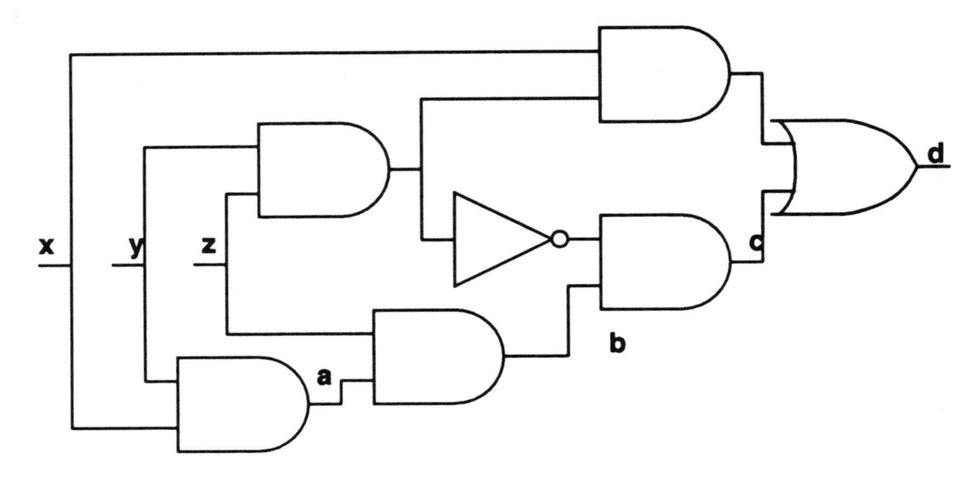

Figure 0.2: A False Path

t_2. The second innovation was to account for the uncertainty in delays in a switching circuit.

What Are We Really After?

Once it is conceded that the temporal behaviour of the circuit in part determines the truth or falsity of a particular path, we must ask whether or not the temporal information that we are given is a perfect model of circuit behaviour. The answer is that it certainly is not.

The real problem is to understand the timing behavior of real circuits manufactured under imprecise control of the manufacturing process, and operating in a variable and loosely controlled environment. We need a criteria under which we can set the clock period so that most of the circuits produced will operate correctly (high yield). Thus we need to analyze not simply a single well-defined circuit (which is hard enough), but a whole family of circuits, characterized by a given fixed topology but with different delays, and generally different circuit parameters. We may be given some statistics giving the probability density functions characterizing the circuit variations, and in general we would like to set the timing of the clocks such that the yield obtained is a compromise between performance and cost considerations. Thus far, no effective procedures exist that solve this problem in its full generality.

One abstraction (and the one used in this book) of this setting, which is frequently used, is to model the delay by its upper bound. Thus for each gate in the circuit, we are given a maximum gate delay which is the slowest time a signal can propagate through the gate for any manufactured circuit operating in any of the allowed environments. We require that the analysis produced gives an upper bound such that any circuit in the circuit family will operate correctly within the time bound derived.

It is interesting to try to understand the difficulty of this problem. Even given the unreasonable assumption that a simulation like SPICE would take almost no time, we are faced with two difficulties. First, what inputs do we give to the SPICE simulation, i.e. how do we stimulate the proposed critical path? Indeed, in order to stimulate a particular path, we may need to simulate a whole sequence of inputs. Second, which circuit of the circuit family do we simulate? After all, during a simulation, the parameters characterizing the circuit are fixed, i.e. we only simulate one circuit of the family. We certainly can't afford to simulate all the circuits of the family, and we have no technique to find the slowest circuit. All of this points to the need for a rigorous mathematical method which can characterize the entire family.

What is done in this book is to derive a criterion characterizing the slowest critical path for the entire family. We show that this criterion can be applied to the "slowest" circuit of the family, i.e. the circuit which has each gate delay set to its largest value. Thus the criterion derived allows us to know which circuit to analyze, overcoming the second problem. Also, the criterion is basically a "symbolic" simulation, overcoming the problem of how to stimulate the circuit.

In general, the numbers we derive are too conservative since we use the worst-case delay through a gate to characterize its delay, thus ignoring other information possibly available about probability distributions. However, within this restriction, we are able to give a criterion which is the tightest one known at present. Indeed, we conjecture that it is the best one can do without requiring a sequence of inputs in the analysis. For example, the viability criterion provides a bound on the delay, it identifies the critical path, and provides a single input vector (not a sequence) which is associated with this path. The single input vector is such that it will stimulate the path in question (during a simulation) provided the internal state of the capacitive nodes of the circuit initially have their most pessimistic logic values. We conjecture that no other criteria can give a better bound without knowledge of the initial internal values on the capacitors. This is only a conjecture and remains to be proved. This conjecture has a troubling consequence, however: we can't guarantee that we can actually stimulate the circuit that we report as the longest. It turns out that this is largely due to the fact that in a logic network, each node can undergo multiple changes in value. As a result, it is extremely difficult to predict what value a node has between the time it undergoes its first and last event. However, in circuits that do not undergo multiple changes in value, or hazards, this restriction is removed. We show that these circuits are simply the set of dynamic MOS circuits, and show that for these circuits precise timing information is available.

We would also like to have a criterion which is fast to compute. In some applications, we would like to identify the critical path, and change it by speeding it up. Having done this we need to identify the next critical path and repeat the process. In such applications we may be willing to be somewhat more conservative in our estimations in order to obtain a faster computation. We have provided a framework from which a spectrum of correct algorithms can be obtained. However, much work remains to be done to determine which options provide the best trade-off between execution speed and precision of the estimate. At one extreme of this spectrum is the longest weighted path in the circuit, which can be computed extremely fast but gives conservative estimates (thus leading to the false path problem). The other end of the spectrum is viability analysis which includes many SAT problems as subproblems. By easing up on the criterion employed and the degree to which the SAT problems are solved, we can obtain a variety of algorithms in between. These various choices remain to be explored thoroughly.

Although much remains to be done to come up with an ultimate set of practical algorithms, the theory and ideas presented in this book have already found useful application. We hope that the increased understanding that this theory provides will help unlock some of the enchanting mysteries about the relation between time and functionality in digital systems.

INTEGRATING FUNCTIONAL AND TEMPORAL DOMAINS IN LOGIC DESIGN

THE FALSE PATH PROBLEM AND ITS IMPLICATIONS

Chapter 1

Introduction

The two classic parameters of integrated circuit design are speed and area. The cost of an integrated circuit is linearly related to the *yield* (that is, to the percentage of instances of the circuit which function correctly). In turn, yield is inversely related to the probability of a fatal defect in the material substrate, which is exponentially related to active area of the circuit. Hence, to a first approximation, the cost of an integrated circuit is a function of the area of the circuit.

Speed and its correct measurement affect both the performance and correctness of an integrated circuit. Performance goes without saying. Correctness follows from the observation that a circuit takes time to settle at a final value. Consider a generic integrated circuit: this consists of networks of combinational logic partitioned by storage elements called *latches* or *registers*. Such latches are typically controlled by a *load* line. When a load line is high, a latch changes state in response to changes on its input. During these periods, the latch is said to be *open*. When the latch is not responsive to changes on its input, the latch is said to be *closed*. If the load line is a clock line (as is typical in conventional designs), the circuit is said to be *synchronous*. Further, one can see that the effective value of the combinational network feeding a latch is the last value on the output of the network before the clock line goes low. Hence, if the correct value of the combinational network is to be computed in response to some input vector, the controlling clock must be high for long enough to permit the circuitry to arrive at a final value. This is called a *timing constraint* on or a *timing specification* of the circuit . A critical question concerning integrated circuits is whether they meet their timing specifications, and answering this question – and coercing circuits to meet their specification – is a major focus of research in computer-aided design.

Measurement of area is trivial at low levels of design, and is easily and accurately estimated at various higher levels through the use of abstract metrics which have been observed to correlate well with final layout area [1]. Speed – or, more precisely, delay – is far harder to measure. At the mask level, the circuit forms a network of transistors. Each transistor, when conducting or "on", acts as a resistor through which the gate on a succeeding transistor can charge or discharge, and so turn on and conduct. Analyses of this form yield a system of ordinary linear differential equations, which in turn may be solved by any number of numerical methods; in particular, the SPICE family of circuit simulators [69] [76] has enjoyed wide popularity over the last 15 years in performing this calculation. More recently, relaxation-based techniques such as RELAX [91] have been introduced to perform this calculation.

Circuit simulation techniques of this form are highly accurate, but have one drawback. Each signal contributes one differential equation to the system. Since circuits of 100,000 or so signals are fairly common, the computation task involved even for the most naive simulation technique (SPICE was originally a backward-Euler method) is herculean. Much recent work has addressed this problem through the use of specialized hardware or massively parallel computers [25] [88], with some success. However, in many CAD environments the use of hardware-intensive solutions is impractical, and software solutions are still much desired.

The software approach to this problem involves dealing with circuits at a higher level of abstraction. Conceptually, circuits may be thought of as networks of discrete components. These components may be arbitrarily large or small, though the utility of timing analyzers which work on large components is problematic, since the delay characterization of such components is usually fairly inaccurate. The most common abstraction is at the level of an atomic boolean function. With each such component, or *gate*, a specific *delay* is associated. In this abstraction, both the waveforms and the static values associated with the various predecessors of the gate are ignored. The circuit is then isomorphic to a weighted, directed graph, where the nodes of the graph are the gates of the circuit

[1] for example, logic synthesis tools estimate area by counting the number of literals which appear in the factored-form description of a circuit

and the weights on the nodes are the delays of the gates. The delay of the circuit is simply the longest path in this graph. Finding this longest path is relatively easy; indeed, if the network is acyclic (as it is in the case of a combinational circuit), the algorithm to find the longest path is the well-known ***topological sort*** procedure [53], which is known to be $O(|V| + |E|)$, where V is the number of nodes in the graph (gates in the circuit) and E is the number of connections between then. Programs of this sort are called ***Timing Verifiers*** or ***Timing Analyzers.***

Timing Analysis is a good idea; and, like other good ideas, it has many parents. The idea of timing analysis dates as far back as the PERT project at IBM, and the original idea to use topological sort for the problem of timing analysis of logic circuits appears to have originated with Kirkpatrick and Clark [51]. Interest was renewed with the advent of the VLSI era in the early 1980's, and research focussed on two major areas. First, the computation of the delay associated with each discrete component (the so-called *delay model*) became a major topic of research; work on delay models was a central focus of the programs CRYSTAL [71] and TV [43]. CRYSTAL also broke circuits down not by logic gate, as was the common practice among timing analyzers, but into units called *stages.* A stage was defined as a path between the gate of a transistor or an output node and a single source. Second, the restriction to combinational (acyclic) circuits of boolean gates was thought too restrictive; both CRYSTAL [71] and TV [43] used event-driven simulators of the sort introduced by Bryant[20] in MOSSIM. In these programs, the transistors were explicitly modelled as bidirectional switches. Other innovations of the period included the introduction of *slacks* (differences between the time a signal was required and the time it arrived) by the TIMING ANALYZER [40].

Early timing analyzers were handicapped by poor delay models. Over the next several years, research continued into both scheduling procedures for non-combinational (i.e., cyclic) networks and into improved delay models. In 1984, Ousterhout [72] contrasted the accuracy of CRYSTAL under a *lumped* vs *slope* delay model. The lumped model (so-called because the capacitance is summed or lumped into a single large capacitor of value C which is presumed to discharge through a similarly-lumped resistor of resistance R, yielding a delay of RC) was shown to yield an error of 25% when compared to a SPICE simulation; using the delay models of Penfield, Rubinstein and Horowitz [78][74], in which a series of linear equations were derived for each delay[2], led to an estimate within 10% of the benchmark SPICE estimate. The relative accuracy of the latter model made the use of CRYSTAL and similar programs attractive for finding the relative ordering of paths in a circuit. Those paths found to be *critical*: those which took the longest to complete, or had the smallest slacks, or both – could subsequently be extracted and simulated in isolation, and the delay estimate refined. Later programs such as E-TV [50] took this approach to its logical conclusion, incorporating the relaxation-based circuit simulator ELOGIC [49] into the program and using the simulator to derive accurate values for the delays down the long paths.

Similarly, in 1987 Bauer, et. al, [5] introduced a new timing analyzer called SUPERCRYSTAL. SUPERCRYSTAL's two distinguishing features were, first, that the waveform over any capacitor in the circuit was approximated by a piecewise exponential waveform, and, second, that the effective resistance across a conducting transistor was determined by the voltage across the transistor, as opposed to being a single number given by the mean. In 1988, an improved version of SUPERCRYSTAL, renamed XPSIM, was announced [4]. XPSIM had been modified to explicitly simulate each stage using the approximate exponential function method [28] with a multirate time step. These improvements led XPSIM to demonstrate SPICE-level accuracy in a fraction of SPICE's runtime, making it suitable for use in timing analysis.

A difficulty with these efforts was that, in general, each timing verifier used either only a single delay model or a small set of models, which was in general only useful for one level of abstraction; timing verification was run at various levels of abstraction, each of which required a different model. In an effort at standardizing and parameterizing earlier work, Wallace and Sequin introduced an abstract version of a timing verifier, a program called ATV[86][87]. ATV's principle attraction was that a user could verify a design at varying levels of abstraction through the selection of parameters to Wallace's single, abstract, model. Further, since many existing models corresponded to a specific selection of such parameters, in some sense ATV represented many timing verifiers in one.

Though the use of accurate delay models has removed one source of systematic inaccuracy in timing verifiers, another remained. The purpose, after all, in discovering the delay down the longest path in a circuit

[2] Actually, the Penfield-Rubinstein model contained a logarithmic term as well as a linear term

is to determine how long a signal travelling down this path will take to reach the terminus. This information is irrelevant if no signal will travel down the circuit. This phenomenon is generically known as the *false path* problem.

The false path problem fundamentally arises because timing verification is *value-independent*; the states of the various wires into a node are ignored, and so presumed to always propagate the value of the preceding node on any path of interest. This is in contrast to simulators, which are value-dependent. Hence, in any *mixed-mode* simulation, where the critical path is identified by timing verifiers and whose length is determined to great accuracy by simulators, an essential problem is to find an input vector which exercises the long path identified by a timing analyzer. This is a particularly acute problem when one is using a simulator capable of simulating an entire circuit, such as XPSIM. If no such vector exists, then the path is said to be *false*.

We draw on the following observation in the analysis of the false path problem. Each node in a circuit can only propagate values from one of its inputs to its output if the other inputs are in a *sensitized* state; in the picture of a network of transistors, that the excitation of the transistor corresponding to the input must open a single conducting path from the output capacitor to ground (power). This forces the other transistors in the network to either unexcited or excited states; if one associates a boolean variable with the control on each transistor, it is easy to see that the set of such states represents a boolean function; this function is a function of the other inputs to the gate, called the *side inputs* to the gate. Indeed, if the excitation, or not, of the relevant transistor forces the output node to discharge or not, one can see that the boolean value represented by the output node is entirely determined by the value of the input control.

At a higher level of abstraction, if one views the circuit as a set of gates, the relevant states of the side inputs may be deduced from the logic function represented by the gate. In this sense, the false path problem is not merely a problem encountered in MOS VLSI designs but in all level-sensitive boolean logics; the scale and complexity of VLSI design makes the problem especially acute, however. Further, as we shall see below, the uncertainties of delay in integrated circuit design make the problem rather more rigid in this technology than in others.

The remainder of this chapter is organized as follows. In section 1.1, we will discuss an abstract picture of a circuit and formulate the circuit timing analysis problem. In section 1.2, we will formally define the false path problem. In section 1.3, we go over some of the notation we'll be using in this book. In section 1.4, we will introduce some notation of modern logic synthesis which will aid in the analysis of the false path phenomenon. Finally, in section 1.5, we outline the remainder of the book.

1.1 Timing Analysis of Circuits

In this section, we formulate the timing analysis problem on circuits as a path-finding problem on weighted graphs. One can picture a circuit as a graph of nodes, each of which computes some function. The choice of node is arbitrary, and represents a trade-off between accuracy and efficiency. A convenient choice is the representation of a *transistor group*[4], which is a generalization of a boolean gate. A transistor group is a maximal collection of transistors and their control connections such that, for every transistor in the region, its control connection lies outside the region. If there are no pass transistors or transmission gates in the design, this definition simplifies to that of a boolean gate. It is this definition that we adopt for the purposes of our discussion here. The edges of the circuit represent the interconnections of modules; since by construction each terminus of an edge represents the control of some transistor, and since signal flow is always to the control of a transistor, each edge in the graph is *directed*. If we assume that the circuit is combinational, as we do in this thesis, the graph is further acyclic.

1.1.1 Delay Models

Once a network of nodes is chosen, delay is conventionally represented by weights on the nodes (or, equivalently, the edges) of the circuit. The derivation of these weights is called the *delay model* of the circuit. Delay models are variously derived, but basically break down into *static* and *dynamic* models. Static models have the property that the delay across each node is statically determined by the graph. The delay across a node is not then a property of the waveform emanating from an input. These are used in HUMMINGBIRD [89], and comprise the simplest (and, perhaps, the most commonly-used) of CRYSTAL's delay models. These can be represented by a graph with numeric weights on the edges.

Dynamic delays, on the other hand, are functions not only of the graph but also of the input waveforms. In general, timing analyzers using dynamic models compute for any node not only the delay across a node, or its arrival time, but also a waveform of the form $V = f(t, I)$, where V is the voltage across the node, t is time, I is an input waveform, and where f is a continuous, monotone function. The "delay" across the node is generally defined as $\{t|V(t) = T\}$, for some threshold value T. Dynamic models vary from very crude (CRYSTAL simply had a table of delays) to highly sophisticated (SUPERCRYSTAL used explicit simulation).

In general, dynamic models are more accurate than static models. Various refinements have been made to the basic static model to improve it. First, it was realized that CMOS gates are composed of dual networks of PMOS and NMOS transistors. Due to differences in electron mobility through the PMOS and NMOS transistors, and/or differences in the length of series chains through these networks, the effective resistance through the PMOS and NMOS sides can be unequal; this is reflected in unequal delays in transitioning the output node from 0 to 1 (the *pullup* transition) than in transitioning the output node from 1 to 0 (the *pull-down* transition). This is represented in the graph by assigning a pair of weights to each node, one in response to a *rising* edge, and one in response to a *falling* edge. This delay model may be thought of as an extremely crude waveform model.

Further enhancements to the static model are possible. In general, the delay response of a gate to one input may be different than that of another; the transistors corresponding to the inputs may be of different sizes, may be driven by differently-sized gates, be attached to nets of varying capacitance, or may appear in different positions in the transistor network that corresponds to the gate. Any of these factors may affect the delay across a node, and so it is natural to separate the delay across a node into delays across each input. This can be modelled by attaching the delays to the incoming edges of a node, not to the node itself[3]. Alternately, one can consider adding to each edge in the graph a node called a *static delay buffer* with the appropriate weight; in this way the delays across the edges can be modelled by delays across nodes in an isomorphic graph.

For convenience, when the theory and algorithms underlying the false path problem are developed in the sequel, a static delay model with one

[3]This is TV's "dynamic" model

delay across each node is assumed. Nevertheless, the results hold for all static delay models unchanged through the isomorphisms developed above. We'll remind the reader of these and develop this theme more fully later.

1.1.2 Graph Theory Formulation

The static timing analysis problem is therefore to find the longest acyclic path in a weighted, directed graph. Consider the special case where the graph is acyclic. The sources of such a graph are called the *primary inputs* of a circuit. Some nodes, including all sinks of the graph, are designated as *primary outputs* of the circuit. We can transform the graph by attaching formal terminal output nodes to each primary output; such a transformation does not affect the timing properties of the circuit if the formal terminals have zero weight, but permit us the convenience of treating the primary outputs and sinks as identical, so, for the remainder of this book, on such graphs the primary outputs are designated as the sinks. Each node n not a primary output has some set of successor nodes in the graph; these are called the *fanouts* of n, and are designated $FO(n)$. Similarly, each node n not a primary output has some set of predecessor nodes in the graph; these are called the *fanins* of n, and are designated $FI(n)$. The transitive closure of FI is called the *transitive fanin* of n, and is denoted $TFI(n)$. The transitive closure of FO is called the *transitive fanout* of n, and is denoted $TFO(n)$. Each node n in the graph has a *level*, denoted $\delta(n)$. $\delta(n)$ is defined as follows:

$$\delta(n) = \begin{cases} 0 & n \text{ is a PI} \\ \max_{p \in FI(n)} \delta(p) + 1 & \text{otherwise} \end{cases}$$

Note that $\delta(n) > \delta(p) \;\; \forall p \in TFI(n)$ and $\delta(n) < \delta(p) \;\; \forall p \in TFO(n)$. The maximum level over all nodes in the graph is called the *diameter* of the graph, and is denoted D.

If the graph is acyclic and the weights are static, the longest-path problem is easily solved. The nodes are ordered by level by a very famous linear-time algorithm, topological sort [79] [53]. The maximum *distance* of a node n from a primary output $D_1(n)$ is thus defined:

$$D_1(n) = \begin{cases} 0 & n \text{ is a sink} \\ \max_{p \in FO(n)} D_1(p) + w(n) & \text{otherwise} \end{cases}$$

and the maximum distance of n from a primary input is defined:

$$D_2(n) = \begin{cases} 0 & n \text{ is a PI} \\ \max\limits_{p \in FI(n)} D_2(p) + w(n) & \text{otherwise} \end{cases}$$

and a longest path is then a sequence of nodes, $\{f_0, ..., f_m\}$ such that

$$D_1(f_i) + D_2(f_i) - w(f_i) = K$$

where:

$$K = \max_{n \text{ is a node in the graph}} D_1(n)$$

If the graph contains cycles, then the problem of finding the longest path is $\mathcal{NP}$-hard [30], and in practice is solved by a technique called switch-level simulation.

Note that if $w(f) = 1$ for each f, we have $D_1(f) = \delta(f)$ for every f.

Since D_1 is the relation conventionally referred to as the delay d of a node, we also write $d(f)$ for $D_1(f)$ for a node f. Since the delay of a node is equal to the length of some path, we also write $d(\{f_0, ..., f_m\})$ for the length of a path.

1.2 The General False Path Problem

Timing verifiers are typically quite fast; indeed, for fully-restoring combinational logic the problem is simply that of finding the longest path through a directed acyclic graph, which is well known to be $O(|V| + |E|)$. However, these programs will always identify the longest path as the critical path of the circuit. This path, however, is not the path of real interest: the path of interest is the longest path down which a signal can propagate. Paths down which no signal can propagate are called *false paths*, and the problem of identifying them, and so finding the longest true path through the circuit, is known as the *false path problem.*

Consider, for example, the circuit in figure 1.1. For x to propagate to a, we must have $y = 1$. For a to propagate to b, we must have $z = 1$. But for b to propagate to c, we must have $y = z = 0$. Hence the path $\{x, a, b, c, d\}$ *appears* to be *false.*

The false path problem has been known for some time. The earliest complete discussion in the literature appears to be due to Hrapcenko [42]

Notation	Definition
$w(n)$	Weight (delay) of node n
$FI(n)$	fanins of node n
$FO(n)$	fanouts of node n
$TFI(n)$	Transitive closure of $FI(n)$
$TFO(n)$	Transitive closure of $FO(n)$
$\{f_0, ..., f_m\}$	path of nodes f_0 to f_m
$d(\{f_0, ..., f_m\})$	$\sum_{i=0}^{m} w(f_i)$
$\delta(n)$	Level of node n in the graph
D	Diameter of the graph (maximum level number)
$D_1(n)$	max distance of node n from a primary output
$D_2(n)$	max distance of node n from a primary input

Table 1.1: Basic Graph Notation

[4]. Hrapcenko demonstrated that, for every integer n, there exists a logic function for which the actual delay of the minimal network is $n + 8$ but for which the longest path is $2n + 8$. Hrapcenko further observed that false paths arise naturally in the design of carry-acceleration adders, and suggested that the longest path through a carry-acceleration adder will be on the order of $2n$ nodes, while the delay will grow approximately as n. This observation correlates well with the experimental evidence of [6], and of this book.

Given the interest in accurate timing verification, considerable importance has been attached to the solution of the false path problem. Early facilities provided for this problem were largely user-oriented, either because the problem was felt to be intrinsically hard or because the authors of the software had an exaggerated respect for designers' intuition. These facilities fell into three major classes.

[4]Hrapcenko's manuscript was kindly brought to the attention of the authors by Prof. N. Pippenger of the University of British Columbia

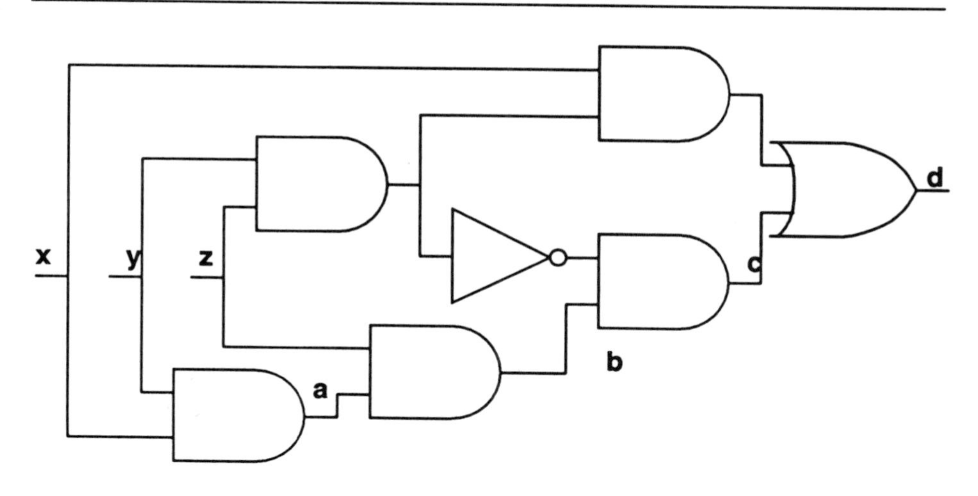

Figure 1.1: A False Path

1.2.1 Explicit Recording of False Paths

Hitchcock's seminal timing analyzer TA [40] contained a facility, called **delay modifiers**, which indicated to TA that, *in the opinion of the designer*, certain paths would never be exercised. This approach was widely adopted; as recently as 1988, newly-reported timing analyzers included such a facility [22]. The difficulty, of course, is that a very large number of paths might eventually be indicated as false; further, there was also the unhappy possibility that the designer's intuition might fail him, though early writers evidently did not feel obliged to discuss this.

1.2.2 Case Analysis

If designer elimination of false paths was tedious, and involved an exhaustive enumeration of many separate paths, then perhaps examining the behaviour of the circuit under assumptions about the input would help. This general technique went under the rubric of *case analysis*.

Ousterhout, in the description of his widely-used switch-level analyzer, CRYSTAL[73] describes the false path problem and this solution technique perhaps as well as anyone in the early literature:

> The value-independent approach is also responsible for the main problem in timing verification. When a timing verifier ignores specific values, it may report critical paths that can never occur under real operating conditions. These false paths tend to camouflage the real problem areas, and may be so numerous that it is computationally infeasible to process them all. In practice, all timing verifiers include a few mechanisms that designers can use to restrict the range of values considered by the program, usually by fixing certain nodes at certain values. This process is called *case analysis*; it is used to provide enough information to the timing verifier to eliminate false critical paths.

Case analysis was not a panacea, however:

> Case analysis must be used with caution. When the user specifies particular values, he restricts the timing verifier from considering certain possibilities; *this may cause critical paths to be overlooked* [italics ours]. Case analysis generally requires several different runs to be made, with different values each run, in order to make sure that all possible states have been examined.

Case analysis was used in both CRYSTAL and in TV. The underlying assumption behind such case analysis is not stated explicitly in either Jouppi's or Ousterhout's work, but the discussions of both, taken together with the technological environments surrounding both projects, make the original motivation clear. Both CRYSTAL and TV were conceived and implemented in co-operation with groups designing VLSI microprocessors [5]. In such processors, which employ two or more non-overlapping (or *underlapped*) clock phases, the most glaringly obvious false paths are those which run through two transparent latches[6] which are active on opposite phases. Hence the original underlying assumption of case analysis was that the signals set to a value would remain constant during the period under evaluation, and, further, that the set of

[5] in Ousterhout's case, the RISC-II/SOAR/SPUR line of processors; in Jouppi's, the MIPS line

[6] A so-called *transparent* latch is a storage element which acts as a pass-through while it is active

nodes whose values are fixed by the assignment of such constant values would have already been set to their assigned values and remain constant throughout evaluation. In other words, case analysis was never designed to account for transient effects during analysis; the false path of figure 1.1, for example, should not be detectable through case analysis, for none of the signals is constant throughout the evaluation period.

Nevertheless, case analysis *can* detect the false path in figure 1.1, by choosing $x = y = 1$; this is an abuse of case analysis, since it violates the underlying assumption of the analysis; namely, that the signal set to a value remains at that value for the entire period of analysis. Nevertheless, case analysis has probably been widely abused in precisely this fashion by designers since the introduction of timing verifiers. The consequences of this abuse we shall see later.

McWilliams, writing before Ousterhout, included case analysis in the SCALD timing verifier[68]. His justification for the use of timing analysis makes clear the underlying assumption that the signals specified by case analysis remain fixed during evaluation of the circuit; however, he permitted signals other than clocks to be specified for the purpose of case analysis:

> If the timing of the circuit never depended on the values of signals, but only on when they were changing or stable, the Timing Verifier would be relatively simple...The signals which are difficult to treat are those whose values affect the circuit timing, and which have different values during different clock cycles. *For example, a control signal which determines whether a register is clocked during a given cycle affects whether the output of the register might change during that cycle.*[italics ours]

Such control signals generally remain stable throughout the clock period; the underlying assumption of case analysis is therefore that the signals set to values during analysis are presumed not to change throughout the period.

1.2.3 Directionality Tags on Pass Transistors

If one considers a network primarily made up of relay-like *pass transistors* (such as, for example, so-called *barrel shifters*), false paths arise from the fact that pass transistors, though nominally bidirectional, are in fact often unidirectional. There are a variety of ways that these can be handled. The CRYSTAL approach relied on user tagging of directionality. Jouppi's analyzer attempted to derive the signal flow direction through a series of rules; his experiments [44] suggest that upwards of 90% of transistors can be correctly categorized by such derivation. SUPERCRYSTAL attaches pass transistors to the nearest stage, and explicitly solves each such stage through the use of a circuit simulator. Cherry, in PEARL adopted both the TV and CRYSTAL approaches, using rules to automatically detect pass transistors; Cherry's ruleset was different from Jouppi's. Regarding the efficacy of the rule-based approach, Cherry notes:

> One circuit that these rules [7] are unable to cope with is a barrel shifter constructed with pass transistors. In general, the only way to determine signal flow in this type of circuit is to know what select nodes[8] are mutually exclusive.

In other words, the complete solution to this problem is derived directly from the solution to the problem outlined in figure 1.1. It is this problem – the problem of finding an automated solution to the false path problem – that is addressed in this book.

The general conclusion that one can draw from this discussion is that the timing analysis of an integrated circuit cannot be made accurately without considering the functional nature of the signals. Further, as we shall see later, the function computed by a signal cannot be accurately determined without taking the timing properties of the circuit into account. The analysis of their interaction is non-trivial, and must be done carefully and correctly in order to solve the false path problem.

1.2.4 Automated Solutions

The key thing about manual, user-oriented solutions to this or to any other problem is that they are in general tedious, and therefore investigation into automated solution is inevitable. As we'll see below, all of the various schemes proposed essentially follow the hand algorithm we executed in finding the false path in figure 1.1. Paths are traced, input to output, and as they are traced values are asserted on signal wires. The

[7] a la TV, but a non-identical ruleset

[8] i.e., which nodes on the gate terminal of the pass transistors

differences in the algorithms lie (partly) in how the paths are traced, and (mostly) in which values are asserted. Choosing the values to assert is called choosing the ***sensitization criterion.*** In the next section, we'll abstract this picture into a functional domain.

1.3 A Note on Notation

This book can be fairly heavy going for those unused to mathematical notation. Over the next couple of sections, we will be introducing the boolean and operator notation that will pervade the book. The goal of this section is tutorial and motivational. First, we hope to persuade the reader that the generality given by the algebraic notation is worth the effort in attempting to understand it; second, we hope to introduce the notation gradually, by example, so that the reader will understand intuitively what, for example, the notation $\mathcal{S}_y \frac{\partial f}{\partial x}$ means for familar gates, such as NANDs and NORs.

The two operators that we'll introduce in the next couple of sections are formalizations and generalizations of simple intuitive ideas. The first, the *boolean difference*, $\frac{\partial f}{\partial x}$, captures what we mean for a node f to be sensitive to one of its inputs, x: that is, when does the value of x completely determine the value of f? The second operator is $\mathcal{S}_x f$, in some sense, equivalent to "ignoring" the value of wire x when computing f.

Let's consider the Boolean difference first. When is f sensitive to x? First, suppose f is an AND gate: $f = g_0 ... x .. g_n$. Then, for f to be sensitive to the value of x, we must have that $g_0 = g_1 = ... = g_n = 1$, for if any $g_j = 0$ then $f = 0$ independent of the value of x. Similarly, if f is an OR gate, $f = g_0 + ... + x + ... + g_n$, then we must have $g_0 = g_1 = ... = g_n = 0$. To cover simple gates, many authors in the testing community refer to *controlling* or *non-controlling* values for gates; a 1 is the non-controlling value for an AND gate, and the controlling value for an OR gate. Similarly, a 0 is the non-controlling value for an OR gate, and the controlling value for an AND gate. The usual condition given in the literature is that, for f to be sensitized to x, then the other inputs to f (called the *side inputs*) to f must be at non-controlling values. This approach has its problems, however. Consider the gate $f = ax + b\overline{x}$. What are the appropriate values of a and b in this case?

The answer begins with the observation that asserting values on signal wires is equivalent to asserting that a certain boolean function is true; for example, asserting that we must have $g_0 = g_1 = ... = g_n = 1$ is equivalent to asserting that the function $g_0 g_1 g_2 ... g_n$ is true. Similarly, asserting that we must have $g_0 = g_1 = ... = g_n = 0$ is equivalent to asserting the function $\bar{g_0}\bar{g_1}...\bar{g_n}$. We can therefore modify the question above to ask, if $f = ax + b\overline{x}$, then what is the appropriate function of a and b to sensitize f to x? Stated differently, what logic function is such that its onset is precisely all the conditions on the side inputs under which f is sensitive to x?

The short answer to this question is that the general condition is that when x toggles, f toggles. As we'll see below, this implies that f evaluated at $x = 0$ is unequal to f evaluated at $x = 1$, or, by formula,

$$f|_{x=0} \oplus f|_{x=1}$$

where $\oplus$ is the EXCLUSIVE OR operator. Now, note that this formula used on the simple gates that we've used to motivate sensitization yields the following. If $f = y_1 ... y_n x_1$, then $f|_{x=0} = 0$, and $f|_{x=1} = y_1 ... y_n = f|_{x=0} \oplus f|_{x=1}$; note that this is equal to the function we derived above. A similar result can be derived when f is an OR, NAND or NOR gate. *Hence, one can view the notation* $\frac{\partial f}{\partial x}$ *as reading "assert a non-controlling value on each side input of* f*".*

The second major operator is the *smoothing* operator, denoted $\mathcal{S}_y f$, and defined as $f|_{y=0} + f|_{y=1}$, where y is an arbitrary variable and f is an arbitrary function. It arises because it is often the case that one wishes to ignore the value of a wire when computing the sensitization function; for example, as we shall see below, Brand and Iyengar [11], when finding the sensitization function for an input x to a gate f, impose an order on the inputs to f, and assert non-controlling values only on the inputs to f that succeed x in the order; the inputs which *precede* x in the standard order are ignored (*smoothed*); no values are asserted on them.

Given that we prefer the functional abstraction given above for sensitization, how can we fit in this practice of ignoring inputs? In particular, how do we ignore the value of y in the expression for $\frac{\partial f}{\partial x}$? Well, if we write $\frac{\partial f}{\partial x}$ out in sum-of-products form, what we operationally would do is simply strike every reference to y or $\bar{y}$ in the expression; a reference to y is equivalent to asserting the value $y = 1$; a reference to $\overline{y}$ is equivalent to asserting the value $y = 0$. Now, functionally, the way we strike every reference to y or $\overline{y}$ from some function g is simply to evaluate g at

$y = 0$, and evaluate g at $y = 1$, and take the OR. This is the definition of the smoothing operator. Hence $\mathcal{S}_y \frac{\partial f}{\partial x}$ may be read "take the boolean difference of f with respect to x and then delete every reference to y", or, more simply, "assert non-controlling values on every side-input to f except y", and, for the purposes of this book, the reader may freely substitute the latter phrase for the formula.

Given that controlling and non-controlling values, and ignoring wires are easier to understand than the functional approach given here, why is it that we insist upon using the functional abstraction? There are two reasons, one of which we've already discussed.

1. The functional approach can be shown to be the generalization of the intuitive approach to networks composed of any arbitrary boolean gates; and

2. Every sensitization criterion has associated with it a function, even if that criterion is described in non-functional terms. However, there is not a great deal of precise semantic content in the intuitive phrase "assert a non-controlling value on every side input to f except y, z, and w", and as a result two separate sensitization criteria are often incomparable in the terms in which they are expressed. However, once it is understood that each such criterion has a boolean function associated with the criterion, it is then possible to calculate the function associated with each criterion and bring the machinery of the boolean calculus to bear in comparing the functions, and thereby comparing the criteria. Essentially, the functional abstraction gives us a rich notational and semantic environment in which to discuss this subject rigorously.

The second point cannot be stressed too much. The fact that the sensitization criteria discussed in this book will be introduced in purely functional terms does not mean that algorithms based on these criteria must explicitly compute these functions. *The fact is, that any algorithm which works by asserting values on selected wires implicitly computes* **some** *function. Similar implicit computation procedures exist for the functions described in the sequel, and these algorithms will be developed in chapter 3.* The functional abstraction is a mathematical convenience from which we are able to prove theorems, not necessarily an operational indicator to those who wish to program the solutions contained herein.

1.3.1 On "Implications"

The name of the game in false path detection is very simple: paths are traced, input to output. As paths are traced, values are asserted on circuit wires. If the values derived on the circuit wires conflict (that is, if some signal x must be both 1 and 0) then the path is judged false. The practice of asserting values on circuit wires is known as deriving implications. We have earlier shown that this practice is equivalent to computing a function; the question we want to answer intuitively here is "what does an 'implication conflict' mean in functional terms?", or, more precisely, "to what formal, functional property does the intuitive phenomenon of 'implication conflict' correspond?"

The answer to this is really very simple. What we seek, after all, is an input stimulus to the circuit so that the path in question will actually be sensitized; i.e., what we seek is an input vector which will force all the implications that we have asserted. Remembering that the implications are really just a way of specifying a logic function, what we're really after is an input vector that forces - or *justifies* - the logic function we've derived to the value 1. In the parlance of boolean analysis, what we want is an input-vector in the on-set of our function, or, equivalently, a vector that satisfies our logic function.

Now, it is clear that no input vector can simultaneously force a wire to both 1 and 0, so any implication conflict represents an unsatisfiable function. The converse is not true, however. Functions do exist which are unsatisfiable but which do not demonstrate any implication conflicts. Hence programs which report a path as false only if they find an implication conflict may report some false paths (by their criterion) as true, but never report true paths as false. This occurs because such programs do not actually verify that an input vector exists which justifies the implications; rather, they just assume that such a vector exists. Programs which actually justify their implications, by contrast, are exact.

The observations we've summarized here permit us to plot algorithms which solve this problem on a two-dimensional graph. The x dimension of the graph is the sensitization criterion: that is, how the implications at each node are derived. The y axis measures how hard the algorithm tries to discover whether the implications are justified or not.

The graph itself appears at the end of chapter 4, and represents the great triumph of the functional abstraction, since it permits programs in this area to be theoretically compared in a comprehensive and organized fashion. Both axes increase in the direction of safety – that is, more conservative estimates of the critical delay. The origin represents the tightest safe criterion – that is, the tightest criterion that does not underestimate the critical delay. The objective of the exercise is then to be the program in the upper-right quadrant closest to the origin. Programs in the lower-left quadrant are guaranteed to underestimate the delay. Programs in the upper-left and lower-right quadrants are aggressive in one dimension and conservative in the other; it is impossible to say whether they under- or overestimate the critical delay on a given circuit, though for any given program in one of these quadrants one can certainly construct a circuit where they underestimate the critical delay; therefore these programs cannot be trusted.

1.3.2 A Final Word on Notation

The summary of this section is easy to state. The results in the rest of this book will be stated in the functional abstraction that we'll derive in the next section. We do this in order to properly model all circuits, and to be able to speak coherently about a variety of different approaches within a single, unified framework. The functional abstraction is a little difficult to read, but the reader should understand that underneath the abstraction something very simple is going on: *all that we are doing is tracing paths forward from the inputs and asserting values on wires. All that the notation tells us is which wires we're asserting values on, and which we're ignoring.* If the reader is ever confused by an expression or notation, we recommend that he consider the effect of the expression on a simple gate: this should greatly aid in understanding the material. In general, we'll try to provide an example after every equation to make understanding easier.

1.4 Logic Notation

In this section, we summarize basic switching theory notation. A summary appears in table 1.2 at the end of this section.

The most common method for representing a logic function is as a sum of terms, $f = t_1 + ... + t_n$, and is read $f = 1$ when $t_i = 1$ for some i. Each term is a product of *literals*; a literal is an ordered pair (v, p) where v is a boolean variable and $p \in \{0, 1\}$ is the *phase* of the variable. The term $t = (v_1, p_1)...(v_m, p_m) = 1$ whenever $v_i = p_i$ for every

i. The notation (v, p) is thought too clumsy, so by abuse of notation the common notation for the positive phase $(v, 1)$ is simply v, and for the negative phase $(v, 0)$ it is simply $\overline{v}$.

$$f = xy\overline{z} + \overline{y}z$$

indicates that $f = 1$ whenever either $x = y = 1$ and $z = 0$ or when $y = 0$ and $z = 1$.

1.4.1 Cubes

In general, one can derive a geometric picture of the boolean n-space as an n-dimensional cube as follows. Consider the n-dimensional Cartesian co-ordinate system. Since each variable can only assume the values 0 and 1, the dimension of the space represented by the variable x can be restricted by planes at $x = 0$ and $x = 1$. Once this has been done in every dimension, the resulting object is an n-dimensional cube.

It is clear that a point in the Boolean n-space is a vertex of the cube; for historical reasons, a vertex is also called a minterm. Moreover, any subspace of the n space is simply another cube, albeit one of smaller dimension. Such a subspace corresponds to a specification of some variables of the space, and hence to a term. Terms are therefore generally referred to as *cubes*[14]. The *size* of a cube is therefore inversely proportional to the number of literals of the term; a term with m literals specifies a cube of size 2^{n-m}.

1.4.2 Cofactors

In much of the sequel we will be discussing the projections of boolean functions on a subspace; this corresponds to a partial evaluation of the function. If one views a function as a set of points on the n-cube (the set of points where $f = 1$), then the cofactor of a function with respect to a literal l, written f_l, is simply the collection of points on the $n - 1$ dimensional cube represented by the literal l. This can be either viewed as a function over this $n - 1$ dimensional space, or (simply by projecting each point on this space onto its neighbour on the $n - 1$ dimensional cube $\overline{l}$) as a function over the boolean $n-$space.

Since the subspace represented by the cube $l_1 l_2$ is the same as that represented by $l_2 l_1$, it is trivial to see that $(f_{l_1})_{l_2} = (f_{l_2})_{l_1}$, and, more generally, f_c is well-defined for any cube c.

Operationally, it is easy to take a cofactor of an m term function over the boolean n–space in time $O(nm)$, assuming some order on the variables. Further, the size of the cofactor of a function in the common function representations (disjunctive and conjunctive normal forms, factored forms, and boolean decision diagrams) is smaller than the size of the original function.

Cofactors derive their importance in logic synthesis due to the following theorem, which is variously credited to Shannon[81] and to Boole:

Theorem 1.4.1 (Shannon Expansion) *For any boolean function f, any variable x,*

$$f = xf_x + \overline{x}f_{\overline{x}}$$

1.4.3 A Family of Operators

In this book, a pair of linear operators over the set of functions on the boolean $n-$ space will be extensively used.

One question that arises in the testing of networks and in the false path problem is the following. Given a function f, and a variable x, what are the assignments to the *remaining* variables such that the value of f is completely determined by the value of x, i.e., $f = x$ or $f = \overline{x}$ (i.e., f changes phase whenever x changes phase; in such a case, we say that f is *sensitized* by x)?

If $f = x$, from the cofactor expansion we must have that:

$$f_x = 1, \quad f_{\overline{x}} = 0$$

i.e., the logic function:

$$f_x \overline{f}_{\overline{x}}$$

must be satisfied. Similarly, if $f = \overline{x}$, we must have that:

$$f_{\overline{x}} \overline{f}_x$$

putting the cases together, we have:

$$\frac{\partial f}{\partial x} = f_x \overline{f}_{\overline{x}} + f_{\overline{x}} \overline{f}_x$$

or, more compactly:

$$\frac{\partial f}{\partial x} = f_x \oplus f_{\overline{x}} \tag{1.1}$$

$\frac{\partial f}{\partial x}$ is described in the testing literature [1] [80], and is referred to there as the *boolean difference.* From that literature, there is the following classic theorem, due to Sellers, et. al. [80].

Theorem 1.4.2 *A network N is testable for a stuck-at fault on node x through output f iff*

$$\frac{\partial f}{\partial x} \not\equiv 0$$

Further, Sellers and his co-workers proved a variety of properties on the boolean difference, which we give without proof here:

$$\frac{\partial^2 f}{\partial y \partial x} = \frac{\partial^2 f}{\partial x \partial y}$$

$$\frac{\partial f+g}{\partial x} = \frac{\partial f}{\partial x} + \frac{\partial g}{\partial x}$$

Another question arises in considering the false path problem. Consider an arbitrary function f, an arbitrary variable x. Under which assignments of the other variables is there an assignment of x s.t. $f = 1$? [9]

Now, we must have $x = 1$ or $x = 0$. If xc is a satisfying assignment of f (i.e., $f(xc) = 1$) then, by the Shannon cofactor expansion, we can say that c is a satisfying assignment of f_x. Similarly, if $\overline{x}c_1$ is a satisfying assignment of f, then c_1 is a satisfying assignment of $f_{\overline{x}}$. Hence, if c is a cube such that either xc or $\overline{x}c$ is a satisfying assignment of f, then c satisfies $f_x + f_{\overline{x}}$. Hence we define the smoothing operator, $\mathcal{S}_x f$ as:

$$\mathcal{S}_x f = f_x + f_{\overline{x}} \tag{1.2}$$

and we have:

Theorem 1.4.3 *Let c be a minterm of $\mathcal{S}_x f$. Then either xc or $\overline{x}c$ is a minterm of f.*

Proof: The discussion above. ∎

Intuitively, we are interested in determining whether some function f is satisfiable, but we have no knowledge as to the value of the variable x. If the satisfiability of the function is dependent upon the value of x, then we clearly will be unable to get a precise answer to this question. Several

[9] this is also called a *satisfying* assignment of f

questions that may be answered precisely arise, which are detailed in appendix B. One is worth detailing here.

Under what assignments of the other variables does there exist a value of x that gives rise to a satisfying assignment of f? Theorem 1.4.3 demonstrates that the answer to this question is the set of satisfying assignments of $\mathcal{S}_x f$. In fact, if we take the answer to the question "Does there exists a satisfying assignment of $\mathcal{S}_x f$?" as the answer to the question "Does there exist a satisfying assignment of f?", then this is an example of a *biased* satisfiability test. It is certainly the case that if there is no satisfying assignment of $\mathcal{S}_x f$, then there is no satisfying assignment of f. However, if x is not an independent variable, as is often the case, then there is a case where there is a satisfying assignment of $\mathcal{S}_x f$ but no satisfying assignment of f (consider the case where $f = xy$ and y implies $\bar{x}$). It is in this context that we will be using $\mathcal{S}_x f$: in the sequel we will be deriving a function that is satisfiable only if a path is true; since we want to reject only false paths, we want to bias this test positively, i.e., smooth out variables whose value is unknown and ask whether such a smoothed function is satisfiable.

Theorem 1.4.4 *Let f, g be any functions, x, y any variables. Then:*

$$
\begin{array}{lrcl}
(i) & \mathcal{S}_x\mathcal{S}_y f & = & \mathcal{S}_y\mathcal{S}_x f \\
(ii) & \mathcal{S}_x(f+g) & = & \mathcal{S}_x(f) + \mathcal{S}_x(g) \\
(iii) & \mathcal{S}_x f & \supseteq & f \\
(iv) & \mathcal{S}_x(fg) & \subseteq & \mathcal{S}_x(f)\mathcal{S}_x(g)
\end{array}
$$

Of these properties, we will be using (i) extensively.

Proof:

$$
\begin{array}{rcl}
(i)\ \ \mathcal{S}_x\mathcal{S}_y f & = & \mathcal{S}_x(f_y + f_{\bar{y}}) \\
& = & (f_y + f_{\bar{y}})_x + (f_y + f_{\bar{y}})_{\bar{x}} \\
& = & f_{yx} + f_{\bar{y}x} + f_{y\bar{x}} + f_{\bar{y}\bar{x}} \\
& = & f_{xy} + f_{\bar{x}y} + f_{x\bar{y}} + f_{\bar{x}\bar{y}} \\
& = & (f_x + f_{\bar{x}_y} + (f_x + f_{\bar{x}})_{\bar{y}}) \\
& = & \mathcal{S}_y(f_x + f_{\bar{x}}) \\
& = & \mathcal{S}_y\mathcal{S}_x f
\end{array}
$$

$$
\begin{array}{rcl}
(ii)\ \ \mathcal{S}_x(f+g) & = & (f+g)_x + (f+g)_{\bar{x}} \\
& = & f_x + g_x + f_{\bar{x}} + g_{\bar{x}} \\
& = & \mathcal{S}_x(f) + \mathcal{S}_x(g)
\end{array}
$$

$$\begin{array}{lrcl}
(iii) & \mathcal{S}_x f & = & f_x + f_{\bar{x}} \\
& & = & x f_x + \bar{x} f_x + x f_{\bar{x}} + \bar{x} f_{\bar{x}} \\
& & = & f + \bar{x} f_x + x f_{\bar{x}} \\
& & \supseteq & f \\
(iv) & \mathcal{S}_x(f)\mathcal{S}_x(g) & = & (f_x + f_{\bar{x}})(g_x + g_{\bar{x}}) \\
& & = & f_x g_x + f_{\bar{x}} g_{\bar{x}} + f_x g_{\bar{x}} + f_{\bar{x}} g_x \\
& & = & \mathcal{S}_x(fg) + f_x g_{\bar{x}} + f_{\bar{x}} g_x \\
& & \supseteq & \mathcal{S}_x(fg)
\end{array}$$

■

By induction on *(i)* we may write, for a set $U = \{x_1, .., x_n\}$ of variables

$$\mathcal{S}_{x_1}\mathcal{S}_{x_2}...\mathcal{S}_{x_n} f = \mathcal{S}_{x_1...x_n} f$$

or, more compactly, as $\mathcal{S}_U f$, and, by induction on *(ii)* and *(iii)*.

$$\begin{array}{rcl}
\mathcal{S}_U(f+g) & = & \mathcal{S}_U(f) + \mathcal{S}_U(g) \\
\mathcal{S}_U(fg) & \subseteq & \mathcal{S}_U(f)\mathcal{S}_U(g)
\end{array}$$

Since the smoothing operator is thus implicitly defined for a set U, it is important to define its behaviour when the set $U = \emptyset$. We choose the obvious definition:

$$\mathcal{S}_\emptyset(f) = f$$

We have two important, though trivial, lemmas on the smoothing operator:

Lemma 1.4.1 *Let V be the set of inputs to a function f, and $U \subseteq V$. For any vector c of the primary inputs, let c_1 be the assignment of the variables in $V - U$ induced by c. Then $c \in \mathcal{S}_U f$ iff there exists some assignment c_2 of the variables in U such that $c_1 c_2$ is a satisfying assignment of f.*

Proof: The only if part is trivial, since $\mathcal{S}_U f \supseteq f$. If part. Induction on $|U|$. If $U = \emptyset$, then $\mathcal{S}_U f = f$, and hence c satisfies f, i.e., the trivial assignment satisfies f. Suppose the statement holds for $|U| < N$. If $|U| = N$, let $U = W + \{x\}$. We can write:

$$\mathcal{S}_U f = \mathcal{S}_x \mathcal{S}_W f$$

We can write the left-hand-side as:

$$(\mathcal{S}_W f)_x + (\mathcal{S}_W f)_{\bar{x}}$$

Notation	Definition
$x(\overline{x})$	Literal representing the value $x = 1$ $(x = 0)$
Cube c	product of literals
f_c	Evaluation of f on the subspace represented by cube c
$\frac{\partial f}{\partial x}$	Boolean Difference of f wrt x $(f_x \oplus f_{\overline{x}})$
$\mathcal{S}_x f$	$f_x + f_{\overline{x}}$
$\mathcal{S}_U f$	$\mathcal{S}_{U-\{x\}}(f_x) + \mathcal{S}_{U-\{x\}}(f_{\overline{x}})$

Table 1.2: Basic Logic Notation

Since c_1 satisfies $\mathcal{S}_U f$, it must satisfy at least one of $(\mathcal{S}_W f)_x$ and $(\mathcal{S}_W f)_{\bar{x}}$. If $(\mathcal{S}_W f)_x$, by induction, there is some vector c_3 satisfying $(\mathcal{S}_W f)_x$, and hence we set $c_2 = xc_3$ and done. Otherwise, there is some vector c_3 satisfying $(\mathcal{S}_W f)_{\bar{x}}$, and hence we set $c_2 = \bar{x}c_3$, and done. ∎

Lemma 1.4.2 *Let U be any set of variables, and c be any cube where every variable in U is set to a value by c. Let $c^* = c - U$, i.e., the variables outside U set to a value by c. Then $(\mathcal{S}_U f)_c = (\mathcal{S}_U f)_{c^*}$.*

Proof: If g is independent of any variable x, then $g_x = g_{\bar{x}} = g$. It follows inductively that if g is independent of the variables set in a cube c, then $g_c = g$. Since $\mathcal{S}_U f$ is a function independent of all the variables in U, it follows that $(\mathcal{S}_U f)_c = \mathcal{S}_U f$. ∎

The smoothing operator and the boolean difference are part of a more general family of such operators, where f_x and $f_{\overline{x}}$ are combined in various ways. For the sake of completeness we detail them in appendix B.

1.5 Outline

The remainder of this book is organized as follows. In chapter 2, we outline the theoretical basis of a correct solution to the false path problem.

The basic problem is that of determining when a path is *true*, or *sensitizable*, and is so referred to as a *sensitization criterion*. In the course of this analysis, we will demonstrate that a criterion which fails to take the *dynamic* nature of the signals in a circuit into account can lead to a timing analyzer which ignores the true critical paths of a circuit, and hence violates the basic *correctness* condition of a timing analyzer. Further, we argue that, since any delay model necessarily overestimates the delay across a node, that a timing analyzer must fulfill a *robustness* condition: namely, it must return an answer that is correct for *every* functionally and topologically identical circuit with identical or possibly lesser delays across individual nodes. We outline a criterion – the *viability* criterion – after demonstrating that the two most obvious criteria are either incorrect or non-robust, and prove that viability is both correct and robust. In chapter 3, we demonstrate that each program in the literature which purports to solve the false path problem is a variant on a single, parameterized algorithm, and that the sensitization criterion is but one of the parameters to this function. We demonstrate that to every criterion corresponds a logic function, and give the logic function both for one of the criteria rejected in chapter 2 and for the viability criterion. We then modify the generic algorithm to correctly compute the viability function. In chapter 4, we explore system considerations. We fully outline the parameter space: in addition to sensitization criterion, the others are the search method, the satisfiability test, and the function representation (the search method is fully developed in chapter 3). We give a general theorem of approximation, a weak – and hence correct and robust, but less tight – version of the viability function, and then demonstrate that two criteria which have appeared in the literature are approximations to weak viability, and so to viability. At an orthogonal axis of approximation, we discuss weak forms of satisfiability. We give experimental results on conjured circuits and on public benchmarks. In chapter 5, we consider hazard-free boolean functions, and show that these are isomorphic to the class of precharged-unate functions. In chapter 6 we show that dynamic sensitization – a sensitization criterion tighter than viability but non-robust on general circuits – is a correct and robust criterion on these circuits. Hence timing analysis on such circuits can yield tighter delay estimates than viability.

The appendices are organized as follows. In appendix A, we examine the complexity of the problem of finding the longest true path by various criteria, and demonstrate that each such problem is a member of the class of $\mathcal{NP}$-complete problems: loosely, the hardest problems whose solution may be verified in polynomial time. In appendix B, we review the family of operators of which the smoothing operator and the boolean difference are the most prominent members, and in appendix C we discuss a fast algorithm for a positively-biased SAT test. In appendix D we review the properties of precharge-unate logic gates.

Chapter 2

The False Path Problem

2.1 Introduction

In this chapter the false path problem is formally treated as a theoretical problem in combinational logic circuits. We begin by reviewing briefly the genesis and practical import of the problem.

Timing analysis and timing optimization of digital circuits is currently recognized as a key area. Optimization requires correct timing behavior, i.e. identification and accurate estimate of a circuit's true critical paths. Typically, critical paths are detected using static timing methods. While these methods are extremely fast, they often lead to serious overestimates of a circuit's delay due to *false paths.* A path is false if it cannot support the propagation of a switching event. In estimating the timing behavior of a circuit, we would like to find the slowest true path.

Several papers have appeared recently in which "false" paths are detected. The classic criterion, most often seen in user-supplied "case analysis" of "incompatible paths" in programs such as CRYSTAL[71] and TV[43], is usually based on *static sensitization.* Under this criterion, a path is false if there exists no input condition such that all gates along the path are sensitized to the value of the previous gate on the path. This approach has recently been formalized in the SLOCOP timing environment[6].

Unfortunately, not every true path is statically sensitizable, so the delay of the longest statically sensitizable path is not necessarily an upper bound on the delay of the circuit. In this chapter we demonstrate that the use of static sensitization as a criterion for the truth or falsity of a path can lead to underestimates of circuit delay, possibly causing the circuit to behave incorrectly.

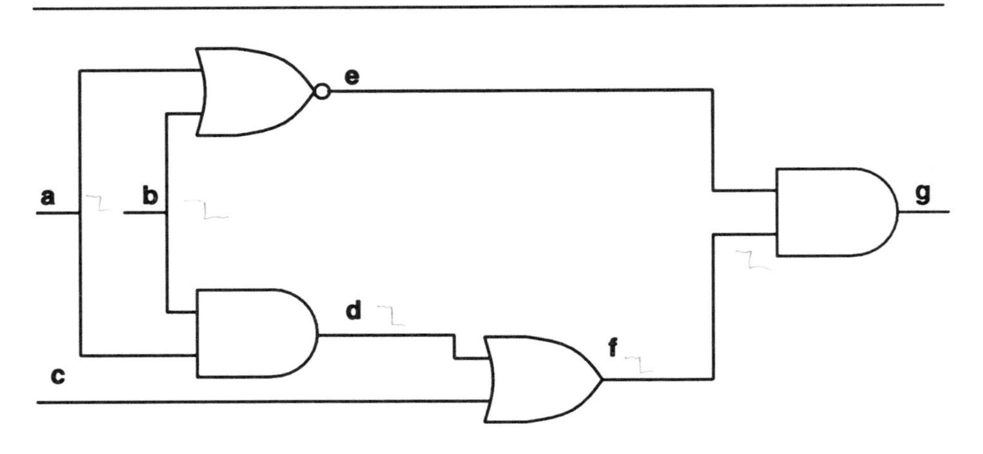

Figure 2.1: A Sensitizable "False" Path

Example: We illustrate this with a small example, taken from [11]. Consider the circuit shown in figure 2.1. Assume that all inputs arrive at $t = 0$, and that the delay on all gates is 1. Consider the path $\{a, d, f, g\}$, of length 3. For a to propagate to d we must have $b = 1$. For f to propagate to g we must have $e = 1$, which implies $a = b = 0$. Hence a static analysis would conclude that this path is false. Similarly, the path $\{b, d, f, g\}$ requires $a = 1$ and $a = 0$, and so a static analysis would conclude that this path is false. Since these are the only paths of length 3, a static analyzer concludes that the longest true path through this circuit is of length at most 2.

Nevertheless, one can see that holding c at 0 while toggling both a and b from 1 to 0 at $t = 0$ forces the output g to switch from high to low at $t = 3$. Hence one of the two paths of length 3 *must* be sensitizable, in the sense that a switching event can travel down it; further, if the clock delay is set to the value (2) of the longest *statically* sensitizable path, the circuit will behave incorrectly on this input. ■

Our first task is the derivation of a criterion for sensitization such that the longest sensitizable path in the circuit is an upper bound for the delay of the circuit.

This task alone, however, is insufficient. Another problem must be dealt with by any algorithm which attempts to compute critical delay. The delay model used in timing analysis methods is a worst case model; it is intended to provide an upper bound for the delay of all circuits which may be manufactured and operated in particular environments. A real circuit is not the idealized circuit of timing models; it is a circuit with the same topology, but with possibly smaller delays at some of the nodes. Hence the estimate provided by the algorithm must hold for an entire *family* of circuits, the "slowest" of which – in the sense of having the slowest components – is typically the one under analysis. In order to use the slowest circuit it is necessary that any critical delay algorithm be *robust* in the following sense: if the delays on some or all gates in the network are reduced, then the critical delay estimate provided by the algorithm is not increased. When the algorithm is applied to the worst-case circuit, a robust criterion thus guarantees that the estimate obtained is valid for any circuit in the family. Colloquially, we refer to this robustness property as the *monotone speedup* property.

In this chapter, we develop a theory which correctly classifies true paths and can be used to provide a correct upper bound for the critical delay in all circuits with the same topology and with equal or less delay at each gate. In section 2.2, we develop our timing model, introduce the concept of event propagation and formally define sensitizable (or true) and critical paths. In section 2.3, the definition of a viable path is given and it is shown that every true path is viable. In section 2.4, it is shown that viability obeys monotone speedup on symmetric networks. In section 2.5, it is shown that every network may be transformed into a symmetric network, and retain all of its viable paths, and thus the longest viable path in the transformed circuit provides the upper bound we seek for the correct timing behaviour. In section 2.6, it is shown that determining the viability of a path is equivalent to computing a logic function.

Thus by finding the longest viable path in the symmetric worst-case delay circuit, we have the correct upper bound required. Of course, the longest path of a circuit also satisfies the above two criteria and so is a correct bound. However, we believe that the longest viable path is a tight bound, although we have not yet been able to show this. We have used our theory to demonstrate that the bounds given in [11, 21, 26] are correct; this theory also demonstrates that those bounds are looser than the bounds developed here. This construction is given in chapter 4.

2.2 Dynamic Timing Analysis

For purposes of clarity, we outline a very simple timing model here. The results of this chapter, however, do not depend on the precise characteristics of this model; we can show that they hold for slope delay models, models with separate rise and fall delays, and different delays on each pin.

Definition 2.2.1 *A* **path** *through a combinational circuit is a sequence of nodes, $\{g_0, ..., g_m\}$, such that the output of g_i is an input of g_{i+1}.*

Definition 2.2.2 *Each node g in a combinational circuit has a* **weight** *$w(g)$. The value of node g at time t is that determined by a static evaluation of the node using the values on its inputs at $t - w(g)$.*

Definition 2.2.3 *We define* **delay** *as follows:*

1. *The* **delay** *through a path $P = \{g_0, ..., g_m\}$ is defined as $d(P) = |\{g_0, ..., g_m\}| = \sum_{i=0}^{m} w(g_i)$. This is also called the* **length** *of the path.*

2. *The* **delay** *at a gate $d(f) = w(f) + max\{d(i)|i \in inputs(f)\}$ for all non-primary inputs f. For all primary inputs x, we define $d(x) = 0$. The weight of a primary input is 0.*

3. *When delays in more than one network are under consideration, the notation $d_N(f)$ denotes the delay at node f in network N.*

The restriction that the delay at the primary inputs is identically 0 is a notational convenience, and does not restrict the body of applicability of this theory. Primary inputs which arrive at $t = T > 0$ can be modelled by assuming that the input arrives at $t = 0$ and that a static delay buffer of weight T is the sole fanout of the primary input. The buffer's fanout is the fanout of the primary input. Primary inputs that arrive ay $t = T < 0$ can be ignored by a simple translation of the time axis.

We assume that the wires of a circuit act as ideal capacitors; that is, once assigned a value the wire holds that value until changed by a computation at its source node. Further, for all negative values of t, the wires of the circuit hold the static values determined by some input vector c_1. Notationally, we capture this assumption by speaking of the value of function f at time t, $f(c_1, c_2, t)$, where c_1 is the input vector from $-\infty \leq t < 0$, and c_2 is the input vector from $0 \leq t \leq \infty$. Clearly we have:

$$f(\neg c_2, t) = f(c_2) \text{ for } t \geq d(f) \tag{2.1}$$

since after time $t = d(f)$, f has assumed its static (final) value.

As we develop the theory, we will use the concept of the "delay" of a function which is not explicitly computed in the network; specifically, of various functions which arise from the boolean difference. These functions do not have a delay within the model developed above. However, it is convenient to assign them a delay. The most reasonable choice is to assume that the computation of these functions is instantaneous. Hence:

Definition 2.2.4 *Let g be any node* not *in a network N. Then $w(g) = 0$.*

Hence, for any such node g,

$$d(g) = \max_{h \text{ is an input of } g} d(h)$$

Definition 2.2.5 *An* **event** *is the transition of a node from a value of 0 to 1, or vice-versa.*

We envision a sequence of events $\{e_0, ..., e_m\}$, each e_i occurring at node f_i, and each event e_i occurring as a direct consequence of event e_{i-1}. We say that event e_0 *propagates* down path $\{f_0, ..., f_m\}$.

Definition 2.2.6 *A path $P = \{f_0, ..., f_m\}$, f_0 a primary input, is* **sensitizable** *if some event e_0 may propagate down this path to the output f_m.*

Definition 2.2.7 *The* **critical path** *of a network is its longest sensitizable path.*

This permits us to consider the boolean conditions for a path to be sensitizable. Let event e_i be the transition of node f_i from 0 to 1. Event e_{i+1} is the transition of f_{i+1} from either 0 to 1 or 1 to 0. In the former case, we have that f_{i+1} tracks f_i, in the latter, f_{i+1} tracks $\overline{f_i}$. The conditions under which this is possible is a boolean function, the arguments of which we call *side inputs*.

Definition 2.2.8 *Let $P = \{f_0, ..., f_m\}$ be a path. The inputs to f_i that are* **not** *f_{i-1} are called the* **side inputs** *to P at f_i. We denote the set of side inputs as $S(f_i, P)$.*

Now, clearly we must have $\frac{\partial f_i}{\partial f_{i-1}} = 1$ when event e_{i-1} is propagated through f_i. We denote the time of event e_i, $t(e_i)$, as τ_i.

Lemma 2.2.1 *Let e_0 propagate down path $\{f_0, ..., f_m\}$, f_0 a primary input. Then $t(e_i) = \sum_{j=0}^{i} w(f_j)$.*

Proof: Induction on i. For $i = 0$, we have e_0 is a change in a primary input and this clearly occurs at $t = 0 = w(f_0)$. Assume for $i \leq j$. For $j+1$, we have that e_{j+1} occurs as a direct consequence of e_j, whence $t(e_{j+1}) = t(e_j) + w(f_{j+1})$, hence $t(e_{j+1}) = \sum_{i=0}^{j+1} w(f_i)$. ■

Theorem 2.2.1 *A path $\{f_0, ..., f_m\}$, f_0 a primary input, is sensitizable iff $\exists$ input vectors $c_1, c_2 \ni \forall i$ $\frac{\partial f_i}{\partial f_{i-1}}(c_1, c_2, \tau_{i-1}) = 1$.*

Proof:
$\Longrightarrow$ $\{f_0, ..., f_m\}$ is sensitizable. By lemma 2.2.1 we have that event e_i occurs at τ_i occurring as a result of the event at τ_{i-1} on f_{i-1}. This requires that either $f_i(c_1, c_2, \tau_i)$ tracks $f_{i-1}(c_1, c_2, \tau_{i-1})$, in which case we must have that $f_{i f_{i-1}}$ is satisfied (so that f_{i-1} going high forces f_i high), and that $\overline{f_{i \overline{f_{i-1}}}}$ is satisfied (so that f_{i-1} going low forces f_i low). Further, the value on f_i at τ_i is statically determined by the values on its inputs at $\tau_i - w(f_i)$, i.e., at τ_{i-1}. This is summarized in the expression:

$$(f_{i f_{i-1}} \overline{f_{i \overline{f_{i-1}}}})(c_1, c_2, \tau_{i-1}) = 1,$$

The other case is that $f_i(c_1, c_2, \tau_i)$ tracks $\overline{f_{i-1}(c_1, c_2, \tau_{i-1})}$, in which case we must have that $f_{i \overline{f_{i-1}}}$ is satisfied (so that f_{i-1} going low forces f_i high), and that $\overline{f_{i f_{i-1}}}$ is satisfied (so that f_{i-1} going high forces f_i low). Further, these conditions must occur τ_{i-1}, as before. This is summarized in the expression:

$$(f_{i \overline{f_{i-1}}} \overline{f_{i f_{i-1}}})(c_1, c_2, \tau_{i-1}) = 1.$$

Putting the cases together, we must have $\frac{\partial f_i}{\partial f_{i-1}}(c_1, c_2, \tau_{i-1}) = 1$, as required.

$\Longleftarrow$ There exist input vectors $c_1, c_2 \ni \frac{\partial f_i}{\partial f_{i-1}}(c_1, c_2, \tau_{i-1}) = 1 \forall i$. Therefore, for every i, we must have that either

$$(f_{i f_{i-1}} \overline{f_{i \overline{f_{i-1}}}})(c_1, c_2, \tau_{i-1}) = 1,$$

in which case $f_i(c_1, c_2, \tau_i) = f_{i-1}(c_1, c_2, \tau_{i-1})$ and the rising (falling) edge which is e_{i-1} is the rising (falling) edge as e_i at τ_i, whence the event e_0 propagates along the path, or

$$(f_{i \overline{f_{i-1}}} \overline{f_{i f_{i-1}}})(c_1, c_2, \tau_{i-1}) = 1,$$

in which case $f_i(c_1, c_2, \tau_i) = \overline{f_{i-1}(c_1, c_2, \tau_{i-1})}$ and the rising (falling) edge which is e_{i-1} is the falling (rising) edge as e_i, at τ_i, whence the event e_0 propagates along the path. These are the only two cases, and in either case e_0 propagates, whence $\{f_0, ..., f_m\}$ is sensitizable. ■ [1]

The basic distinction between the current theory and the previous attempts may now be made clear. The previous theory required that $\frac{\partial f_i}{\partial f_{i-1}}(c_2) = \frac{\partial f_i}{\partial f_{i-1}}(c_1, c_2, \infty) = 1$, a much stronger condition than $\frac{\partial f_i}{\partial f_{i-1}}(c_1, c_2, \tau_{i-1}) = 1$. Paths for which the former condition holds are called *statically sensitizable*. Paths for which the latter condition holds are called *dynamically sensitizable*, or (in light of theorem 2.2.1) *sensitizable*. It is easy to show that:

Theorem 2.2.2 *Every statically sensitizable path is sensitizable.*

Proof: A path is statically sensitizable iff, for all i, $\frac{\partial f_i}{\partial f_{i-1}}(c) = 1$ for some c. Clearly then $\frac{\partial f_i}{\partial f_{i-1}}(c, c, t) = 1 \ \forall t \geq 0$, giving the result. ■

Remark: This proof is obviously trivial in the sense that applying identical vectors will not generate any event to propagate. However, it is clear that if the cube c does not specify f_0, then one can obtain c_1 and c_2 by toggling the f_0 bit.

Remark: Note that the converse to this theorem is false: not all sensitizable paths are statically sensitizable. Indeed, by appropriate adjustment of the internal delays it appears that one can make almost any path in any circuit sensitizable (of course, the timing characteristics of

[1] In the case where the value of $\frac{\partial f_i}{\partial f_{i-1}}$ is changing at τ_{i-1}, to be conservative, we choose $\frac{\partial f_i}{\partial f_{i-1}} = 1$ at τ_{i-1}.

such adjusted circuits vary considerably). Moreover, one can demonstrate fully-testable circuits whose longest dynamically sensitizable path is not statically sensitizable; this statement demonstrates that static vs dynamic sensitizability can be an issue in the timing verification of non-contrived circuits. Indeed, in the circuit of figure 2.1, though the connections of both a and b to the AND gate are non-testable, and d is untestable for stuck-at-zero, the circuit is made fully testable through the addition of a second output, as shown in figure 2.2.

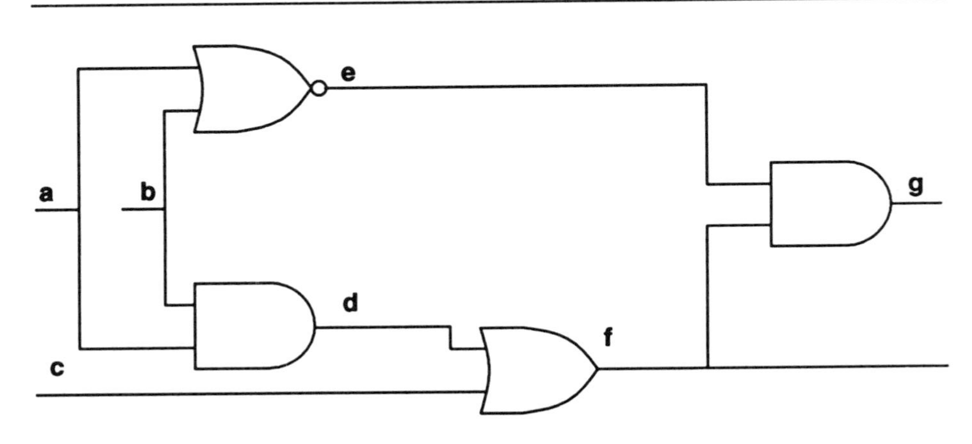

Figure 2.2: A Fully Testable Example

Algorithms which attempt to discover whether a given path is sensitizable must determine whether or not input vectors c_1, c_2 satisfying theorem 2.2.1 exist. There is a wide range of freedom permitted these vectors. However, we may say immediately:

Theorem 2.2.3 *Let $\{f_0, ..., f_m\}$ be a sensitizable path. If $d(\frac{\partial f_i}{\partial f_{i-1}}) \leq \tau_{i-1}$ for some i, then $\frac{\partial f_i}{\partial f_{i-1}}(c_2) = 1$*

Proof: Since $\{f_0, ..., f_m\}$ is sensitizable, $\frac{\partial f_i}{\partial f_{i-1}}(c_1, c_2, \tau_{i-1}) = 1$. But since $\tau_{i-1} \geq d(\frac{\partial f_i}{\partial f_{i-1}})$, we have $\frac{\partial f_i}{\partial f_{i-1}}(c_1, c_2, \tau_{i-1}) = \frac{\partial f_i}{\partial f_{i-1}}(c_2)$ by equation 2.1, whence the result. ∎

Corollary 2.2.4 *The longest path in a circuit is sensitizable iff it is statically sensitizable.*

Proof: The if part is given by theorem 2.2.2. For the converse, observe that the premise of theorem 2.2.3 holds for every f_i on the longest path ■

Recall that every valid criterion must meet the monotone speedup property: if the delays on some or all gates in the network are reduced, then the critical delay estimate for the network is not increased. This guarantee cannot be given by the dynamic sensitization criterion, because the sensitizability of a path is inherently determined by the precise internal delays of the circuit. Hence, one can speed up a circuit and thus make a previously-unsensitizable path sensitizable. This path may be arbitrarily long (though not the longest in the circuit if such is unique); in particular, it may be longer than the longest-sensitizable path in the slower network.

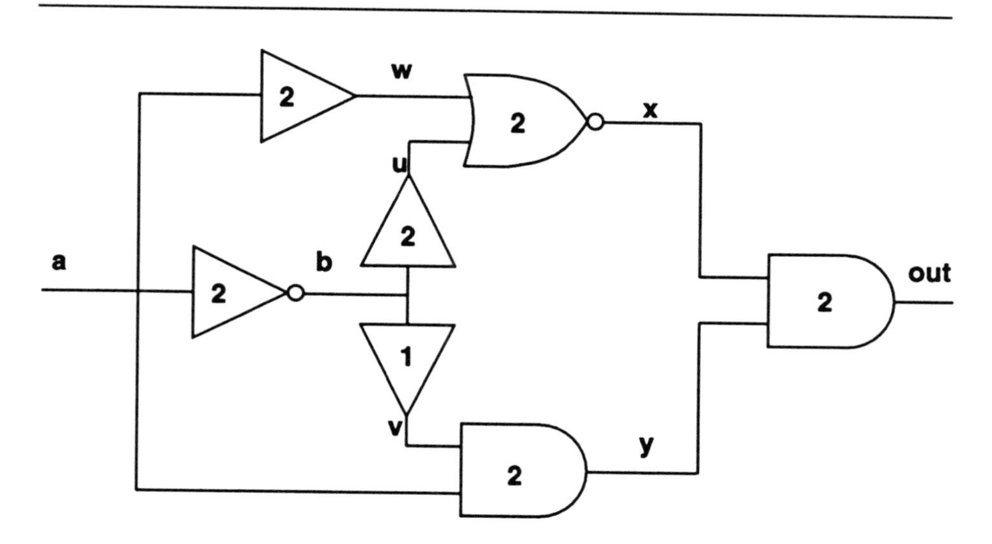

Figure 2.3: Monotone Speedup Failure

An example which illustrates this phenomenon is detailed below.

Example: Consider the single-input circuit in figure 2.3. Assume the delay on all gates are as marked. Note when a is toggled from 1 to 0 at $t = 0$, from $t = 2$ to $t = 3$ there is a 0 on both u and w, so $x = 1$ from $t = 4$ to $t = 5$. However, in this case $y = 0$ throughout, so $out = 0$ throughout. Similarly, when a toggles from 0 to 1, from $t = 0$ to $t = 2$ there is a 1 on each input of y, so $y = 1$ from $t = 2$ to $t = 4$. However, in this case $x = 0$ throughout, so $out = 0$ throughout. This circuit therefore has no dynamically sensitizable paths and its delay is 0.

If we now speed the circuit up by removing the delay buffer between b and u, so that u now arrives at $t = 1$, but all other delays are unchanged, when a is toggled from 0 to 1 we have a zero on each input to x from $t = 1$ (when u turns from 1 to 0) to $t = 2$ (when w turns from 0 to 1). Hence $x = 1$ from $t = 3$ to $t = 4$. But $y = 1$ from $t = 2$ to $t = 4$, so $out = 1$ from $t = 5$ to $t = 6$. Hence there is at least one dynamically sensitizable path in this circuit of length 6; by reducing the delay on the wire from the inverter to u from 2 to 0, we have increased the critical delay on this circuit from 0 to 6. A full timing diagram of the situation appears in figure 2.4. In this diagram, the solid lines represent the behaviour of the "slow" original circuit; the dotted lines the behaviour in the sped-up, or "fast" circuit. ■

This phenomenon – that one can demonstrate circuits where the longest true path of a circuit increases length as components are sped up – appears to hold in every level-sensitive logic where each wire holds its value until the value is changed. In fact, given that the exact delay times at the nodes in a circuit are only determined up to some given tolerance, the sensitizability of a path within a given circuit may vary between two "identical" but separate realizations. *Hence the longest sensitizable path appears to be an inherently nondeterminate property of logic circuits.*

2.3 Viable Paths

Since the longest sensitizable path does not satisfy monotone speedup, we cannot use this criterion to derive a correct upper bound on our family of circuits. We attempt to find a condition on circuits weaker than dynamic sensitization but one that is as strong as possible, certainly tighter than that given by a simple longest-path procedure. The condition C that we seek must possess two properties:

- Every sensitizable path must satisfy C

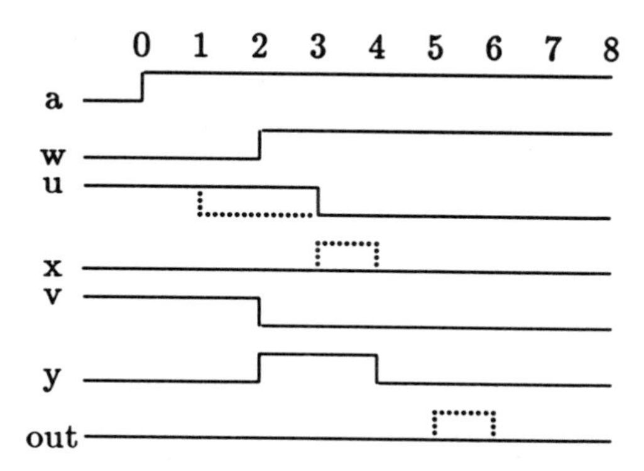

Figure 2.4: Timing Diagram of Monotone Speedup Failure

- $\mathcal{C}$ must satisfy the monotonic speedup property; if network N' is obtained from N by reducing some or all delays, then the longest path satisfying $\mathcal{C}$ in N' must be no longer than the longest path satisfying $\mathcal{C}$ in N

One property that satisfies these constraints is simple longest path. However, this is too weak a condition, and we can do better. A strong property that satisfies these constraints is *viability*. Before we formally introduce the concept of viability, we wish to introduce its motivation.

Fundamentally, a node f_i is dynamically sensitized to an input f_{i-1} at τ_{i-1} but not statically sensitized to f_{i-1} only if the value of the function $\frac{\partial f_i}{\partial f_{i-1}}$ changes value at τ_{i-1} or later. This can only occur if there are events on some set of inputs to $\frac{\partial f_i}{\partial f_{i-1}}$ at or after τ_{i-1}; these are called *late side inputs*. Under these conditions, we may assume that each of these inputs are at *any* value at τ_{i-1}, and hence (to be conservative), we assume that they are set to any value which will propagate the event. Mathematically, we do this by "smoothing" the function $\frac{\partial f_i}{\partial f_{i-1}}$ over the late inputs (see 1.2).

Definition 2.3.1 *Consider a path $P = \{f_0, ..., f_m\}$. Q is said to be a* **side path** *of P at f_i if Q terminates in g, a side input to P at f_i.*

Definition 2.3.2 *A path $P = \{f_0, ..., f_m\}$ is said to be* **viable** *under an input cube c if, at each node f_i there exists a (possibly empty) set of side inputs $U = \{g_1, ..., g_n\}$ to P at f_i, such that, for each j,*

1. *g_j is the terminus of a path Q_j,*

2. *$d(Q_j) \geq \tau_{i-1}$ and Q_j is viable under c*

3. *$(\mathcal{S}_U \frac{\partial f_i}{\partial f_{i-1}})(c) = 1$*

Intuitively, at each node we find the conditions which simultaneously permit a set of side inputs U (a subset of the *late side inputs*) to undergo events later than τ_{i-1}, and the remaining side inputs to statically sensitize the node. It is important to note that this can only occur if there is some assignment to the variables in U which statically sensitizes the node; the effect of the smoothing operator is to permit this assignment to be made, independent of conditions elsewhere in the network. Effectively, the variables in U are made independent variables by the smoothing operator. Note that the case $U = \emptyset$ corresponds to static sensitization.

The intuition behind smoothing off late side inputs may be grasped by considering the case where f_i is an AND gate, $f_i = f_{i-1}a_1...a_n$. In this case, $\frac{\partial f_i}{\partial f_{i-1}} = a_1...a_n$. If the inputs a_i and a_j are smoothed off the boolean difference, however, the resulting expression is

$$a_1...a_{i-1}a_{i+1}...a_{j-1}a_{j+1}...a_n$$

We demonstrate that the criterion of viability under a cube has the two properties we seek; namely, it is weaker than sensitizability, and it has the monotone speedup property. We do so by induction on the maximum distance (in nodes, not node weights) of a node n from the primary inputs, denoted $\delta(n)$, and called the *level* of n. Note for every node m in the transitive fanin of n, we have that $\delta(n) > \delta(m)$, and that the only nodes p for which $\delta(p) = 0$ are the primary inputs. Hence inductive proofs on $\delta(m)$ are really proofs on the structure of a graph; we will be showing that, given that a property holds for each node in the transitive fanin of some node, then it holds at that node.

We check our two conditions, first checking that every sensitizable path is viable.

Theorem 2.3.1 *Let $P = \{f_0, ..., f_m\}$ be a path. If P is sensitizable with $\frac{\partial f_i}{\partial f_{i-1}}(c_1, c_2, \tau_{i-1}) = 1 \;\forall i$, then P is viable under c_2.*

Proof: We prove by induction on $\delta(f_m)$. If f_m is a primary input then trivial. So suppose true for all paths terminating in some f_m such that $\delta(f_m) < L$. Now, consider a path terminating in f_m such that $\delta(f_m) = L$. Let $\frac{\partial f_i}{\partial f_{i-1}}(c_2) = 0$ for some f_i $i \leq m$. (if no such i exists, then the path is viable under c_2 by (1) of the definition, and done). Now, since $\frac{\partial f_i}{\partial f_{i-1}}(c_1, c_2, \tau_{i-1}) = 1$, and $\frac{\partial f_i}{\partial f_{i-1}}(c_1, c_2, \infty) = 0$, there were events on inputs to $\frac{\partial f_i}{\partial f_{i-1}}$ at some $t \geq \tau_{i-1}$. The inputs to $\frac{\partial f_i}{\partial f_{i-1}}$ are the side inputs to P at f_i. Let $U = \{g_1, ..., g_n\}$ be the side inputs where the events occurred. Each event propagated under c_2 to g_j from some primary input h_{0j}, whence $Q_j = \{h_{0j}, h_{1j}, ..., g_j\}$ is a sensitizable path and $d(Q_j) > \tau_{i-1}$. Further, $\delta(g_j) < \delta(f_m) = L$, and hence by the inductive assumption Q_j is viable under c_2. Finally, the side inputs where no events occurred after τ_{i-1} is the set $S(f_i, P) - U$. These are precisely the inputs to $\mathcal{S}_U \frac{\partial f_i}{\partial f_{i-1}}$, whence we must have $\mathcal{S}_U \frac{\partial f_i}{\partial f_{i-1}}(c_2) = \mathcal{S}_U \frac{\partial f_i}{\partial f_{i-1}}(c_1, c_2, \infty) = \mathcal{S}_U \frac{\partial f_i}{\partial f_{i-1}}(c_1, c_2, \tau_{i-1})$. Since $f(c) = 1 \Rightarrow \mathcal{S}_x f(c) = 1$, and since $\frac{\partial f_i}{\partial f_{i-1}}(c_1, c_2, \tau_{i-1}) = 1$, we have that $\mathcal{S}_U \frac{\partial f_i}{\partial f_{i-1}}(c_1, c_2, \tau_{i-1}) = 1$, whence $\mathcal{S}_U \frac{\partial f_i}{\partial f_{i-1}}(c_2) = 1$ and done. ■

The converse to this theorem is false; not every viable path is sensitizable. Clearly the converse cannot hold since viability is robust and dynamic sensitization is not robust.

2.4 Symmetric Networks and Monotonicity

Viability does not possess the monotone speedup property on general networks; however, it *does* possess this property on networks composed of symmetric gates. The objective of this section is to prove this. The proof must be approached indirectly, for the set of viable paths changes as one changes the internal delays of the network. Hence the proof of the monotonicity theorem for symmetric networks is given by a construction: if N' is obtained from N by reducing some delays, and if P' is a viable path in N', then we construct a viable path P in N with $d_N(P) \geq$

$d_{N'}(P')$. Thus N always contains a viable path at least as long as the longest viable path in N'. Having done this, in the sequel we shall show how to apply this result to networks containing asymmetric gates.

Definition 2.4.1 *A function f is said to be* **symmetric** *in some set of variables U if, for every permutation of U, there exists a phase assignment to the variables in U such that f is invariant.*

Example: f is symmetric in the variables x, y if one of the following holds:

$$\begin{aligned} f(..., x, y, ...) &= f(..., y, x, ...) \\ f(..., x, y, ...) &= f(..., \overline{y}, x, ...) \\ f(..., x, y, ...) &= f(..., y, \overline{x}, ...) \\ f(..., x, y, ...) &= f(..., \overline{y}, \overline{x}, ...) \end{aligned}$$

■

Example: $f(x, y) = x + \overline{y}$ is symmetric in x and y, since $f(\overline{y}, \overline{x}) = \overline{y} + \overline{\overline{x}} = f(x, y)$ ■

Lemma 2.4.1 *If f is symmetric in a set of variables U then for every $V \subseteq U$ where $|V| \geq 2$ and $x, y \in V$:*

$$\mathcal{S}_{V-\{y\}}\frac{\partial f}{\partial y} = \mathcal{S}_{V-\{x\}}\frac{\partial f}{\partial x}$$

Proof: Without loss of generality, assume that each variable is assigned the positive phase in the phase assignment of the symmetry definition. If $|U| = 2$, and since $f(..., x, y, ...) = f(..., y, x, ...)$, we have $f_x(y) = f_y(x)$, $f_{\overline{x}}(y) = f_{\overline{y}}(x)$, $\overline{f}_x(y) = \overline{f}_y(x)$, $\overline{f}_{\overline{x}}(y) = \overline{f}_{\overline{y}}(x)$, whence we have $\frac{\partial f}{\partial y}(x) = \frac{\partial f}{\partial x}(y)$ and so:

$$\mathcal{S}_x\frac{\partial f}{\partial y} = \mathcal{S}_y\frac{\partial f}{\partial x}$$

which gives us the result. Now Let $|U| = L$. Consider any $V \subset U$. We have:

$$\begin{aligned} \mathcal{S}_{U-\{y\}}\frac{\partial f}{\partial y} &= \mathcal{S}_{U-\{x,y\}}\mathcal{S}_x\frac{\partial f}{\partial y} \\ &= \mathcal{S}_{U-\{x,y\}}\mathcal{S}_y\frac{\partial f}{\partial x} \text{ (from the base case)} \\ &= \mathcal{S}_{U-\{x\}}\frac{\partial f}{\partial x} \end{aligned}$$

■

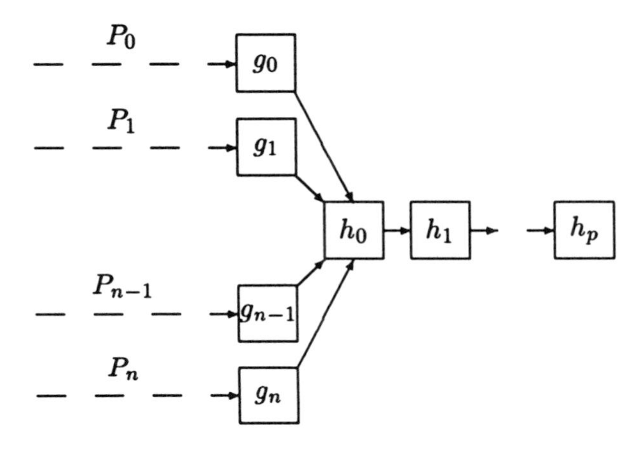

Figure 2.5: "Viable Fork" Lemma

Symmetry is important because we can show monotone speedup for networks composed of symmetric nodes. Further, as is evident from the definition above, most networks are largely symmetric, and thus this theorem has some practical importance.

For proving monotone speedup, we need a technical lemma concerning the existence of viable paths when presented with a set of viable partial paths which conjoin. This is presented in the "viable fork" lemma. Reference to the diagram in figure 2.5 is helpful when analyzing this situation.

Lemma 2.4.2 (Viable Fork) *Let $V = \{g_0, ..., g_n\}$ be a subset of the inputs to some node h_0. Let each g_i be the terminus of a path P_i, a viable path under c. Let $c \subseteq \mathcal{S}_{V-\{g_i\}}\frac{\partial h_0}{\partial g_i}$ for each i. Let $Q = \{h_0, ..., h_p\}$ be any path, such that $c \subseteq \frac{\partial h_{j+1}}{\partial h_j}$ for every j. Then for some i, $\{P_i, Q\}$ is a viable path under c.*

Proof: Since $c \subseteq \frac{\partial h_{i+1}}{\partial h_i}$ for every i, all that must be shown is the viability under c of one of $\{P_i, h_0\}$. This is trivial if h_0 is statically sensitized for any of the g_i by c, so assume not. For some i, $d(P_i)$ is minimal among the P_i, and since by assumption $c \subseteq \mathcal{S}_{V-g_i}\frac{\partial h_0}{\partial g_i}$, and since for each g_j, P_j is viable under c with $d(P_j) \geq d(P_i)$ we have that $\{P_i, h_0\}$ is viable under c by the definition of viability. ∎

In the monotone speedup theorem, we will be conjoining various partial paths which are known to be viable under some cube c onto the common "tail", in this lemma given by $\{h_1, ..., h_p\}$, and we will want to show that one of the resulting paths is viable under c, whence at least one path is viable under c.

Note that by lemma 2.4.1, for symmetric h_0, any V, $c \subseteq \mathcal{S}_{V-g_i}\frac{\partial h_0}{\partial g_i}$ for some $g_i \in V$ iff $c \subseteq \mathcal{S}_{V-g_i}\frac{\partial h_0}{\partial g_i}$ for each $g_i \in V$.

We now turn to the main theorem of this section, which demonstrates that viability has the monotone speedup property in symmetric networks. We proceed in this proof as follows: given a path P' viable under c in a "fast" network N', we demonstrate the existence of a slower path P in the "slow" network N viable under c.[2]

In the proof, the diagram in figure 2.6 is helpful.

Theorem 2.4.1 *Let N' be any network obtained from a symmetric network N by reducing some internal delays. For every viable path $P' = \{f_0, ..., f_m\}$ in N', $\exists$ $P = \{k_0, ..., f_m\}$ a viable path in N with $d(P) \geq d(P')$.*

Proof: Let $\{f_0, ..., f_m\}$ be a viable path in N'. We proceed by induction on $\delta(f_m)$. The base case is trivial, so assume for $\delta(f_m) < L$. Consider the case $\delta(f_m) = L$. If $\{f_0, ..., f_m\}$ is statically sensitizable under c, then done, since this path is viable in every network in the family, and so in N. If not, let f_i be the last node that is not statically sensitized to f_{i-1} by c. Since $\frac{\partial f_i}{\partial f_{i-1}}$ is not satisfied by c, then since $\{f_0, ..., f_m\}$ is viable in N' by the definition of viability there exists a set $U = \{g_0, ..., g_n\}$ of the side inputs such that for each g_j there is a P'_j viable under c in N' with $d_{N'}(P'_j) \geq d_{N'}(\{f_0, ..., f_{i-1}\})$ and with $c \subseteq \mathcal{S}_U\frac{\partial f_i}{\partial f_{i-1}}$. Now, since $\delta(g_j) < L$, by the induction hypothesis for each j $\exists$ a P_j, viable under c in N and terminating in g_j, with $d_N(P_j) \geq d_{N'}(P'_j)$. Further, since $\{f_0, ..., f_{i-1}\}$ is viable under c in N', and $\delta(f_{i-1}) < L$, by induction there is a path $\{a_0, ..., a_k, f_{i-1}\}$, viable under c in N, with

[2] By "fast" and "slow" here we mean that N' has been obtained by reducing some delays in N

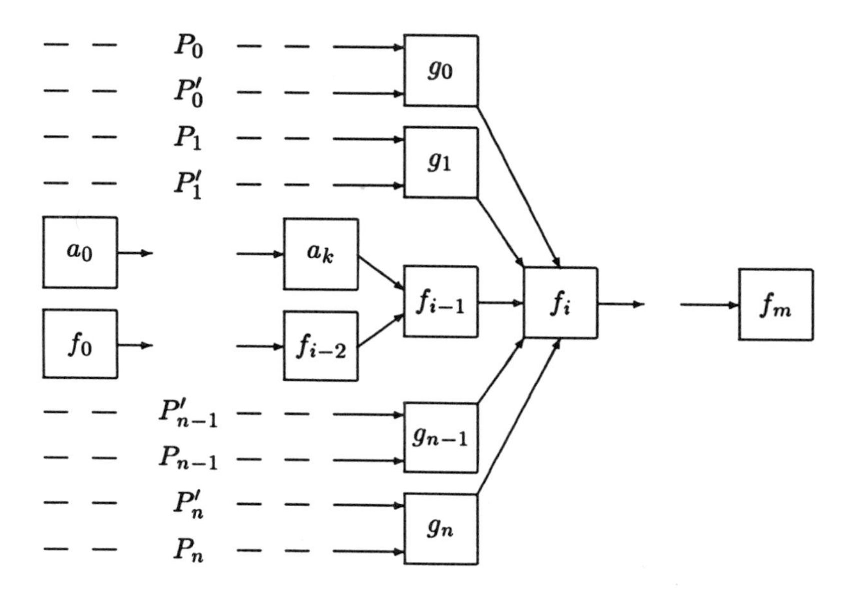

Figure 2.6: Viable Paths in N and N'

$d_N(\{a_0, ..., a_k, f_{i-1}\}) \geq d_{N'}(\{f_0, ..., f_{i-1}\})$. Since N is symmetric, and $c \subseteq \mathcal{S}_U \frac{\partial f_i}{\partial f_{i-1}}$ the set of inputs $U \cup \{f_{i-1}\}$ satisfies the assumptions for the set V of lemma 2.4.2, with $\{h_0, ..., h_p\} = \{f_i, ..., f_m\}$. Therefore one of the $\{P_j, f_i, ..., f_m\}$is viable under c in N, or $\{a_0, ..., a_k, f_{i-1}, f_i, ..., f_m\}$ is viable under c in N.

$$
\begin{aligned}
d_N(\{a_0, ..., a_k, f_{i-1}, f_i, ..., f_m\}) &= d_N(\{a_0, ..., a_k, f_{i-1}\}) + d_N(\{f_i, ..., f_m\}) \\
&\geq d_{N'}(\{f_0, ..., f_{i-1}\}) + d_{N'}(\{f_i, ..., f_m\}) \\
&\geq d_{N'}(\{f_0, ..., f_{i-1}, f_i, ..., f_m\})
\end{aligned}
$$

and:

$$
d_N(\{P_j, f_i, ..., f_m\}) \geq d_{N'}(P'_j) + d_N(\{f_i, ..., f_m\})
$$

$$\geq d_{N'}(\{f_0, ..., f_{i-1}\}) + d_{N'}(\{f_i, ..., f_m\})$$
$$\geq d_{N'}(\{f_0, ..., f_{i-1}, f_i, ..., f_m\})$$

So all paths have greater delays in N than the path $\{f_0, ..., f_m\}$ in N', and since one must be viable under c in N, we are done. ■

One might wonder at the utility of this theorem, since it is easy to exhibit asymmetric gates: the gate $xy + z$ is clearly symmetric only in x and y. However, we can transform any network of asymmetric gates into a symmetric network, at some increase in the number of viable paths. We develop this in the next section.

2.5 Viability Under Network Transformations

It is well-known that any boolean function may be written in sum-of-products form. Now, we may consider any gate in an arbitrary network to be implemented in this way: for each term, there is a single *and* gate realizing the term, and the *and* gates are the inputs to a single *or* gate which realizes the function. Now, if we assign the internal *and* gates to have weight 0, and the *or* gate realizing node f to have weight $w(f)$, we will not have changed the timing characteristics of the network in any way. Our purpose in this section is to show that the set of viable paths of the network is not decreased by this transformation. In practice, the *and/or* transform is one of a large class of such transforms, which we call *macroexpansion* transforms (since each gate is macroexpanded). Formally, we can write the definition this way:

Definition 2.5.1 *Let N be a network of arbitrary gates. A transform $\mathcal{T}$ is called a* **macroexpansion** *of N if $\mathcal{T}(N)$ has the properties:*

1. *Each gate in $\mathcal{T}(N)$ belongs to precisely one subnetwork $\mathcal{T}(f_i)$.*

2. *For each $f_i \in N$, $\mathcal{T}(f_i)$ is an acyclic digraph consisting of zero or more* **internal** *nodes, each of which has weight 0 and whose fanouts are nodes of $\mathcal{T}(f_i)$, and one output node, designated $\mathcal{O}(f_i)$, whose weight is $w(f_i)$ and whose fanouts are the fanouts of f_i.*

3. *For each $f_i \in N$, $\mathcal{T}(f_i)$ realizes the logic function f_i.*

Informally, each gate in the network is replaced by a subnetwork implementing it.

The example of the *and/or* transform is instructive. If $\mathcal{T}$ is the *and/or* transform, and if $f = c_1 + c_2 + ... + c_n$, each c_i a cube, then $\mathcal{O}(f)$ is an *or* gate whose inputs consist of n *and* gates, $\mathcal{T}(c_1), ..., \mathcal{T}(c_n)$. $w(\mathcal{T}(c_i)) = 0 \ \forall\, i$, $w(\mathcal{O}(f)) = w(f)$.

Our purpose is to show for any generic macroexpansion transform $\mathcal{T}$, and for every viable path P in N, there is at least one corresponding viable path $\mathcal{T}(P)$ in $\mathcal{T}(N)$, with $d_N(P) = d_{\mathcal{T}(N)}(\mathcal{T}(P))$. This shows immediately that the critical path delay returned by the viable path algorithm on a macroexpanded network is an upper bound on the true critical path delay. Moreover, since we can certainly macroexpand any network into a symmetric network, we can apply the monotonicity theorems to the symmetric network and be assured that the critical delay obtained is not only upper-bounded in the network N, but also in any faster network N'.

This theorem is a little difficult, and rests heavily on the relationship between the boolean difference, static sensitization, and testing, and on the properties of the smoothing operator. First, note that since the operations of cofactoring and complementation are independent of the implementation of a function, then so to is the boolean difference. Hence we have that

$$\frac{\partial \mathcal{O}(f_m)}{\partial \mathcal{O}(f_{m-1})} = \frac{\partial f_m}{\partial f_{m-1}}$$

The next piece of this puzzle comes from a technical lemma. We wish to show that for every gate in the macroexpanded network, if c satisfies $\mathcal{S}_U \frac{\partial f_m}{\partial f_{m-1}}$, then there is a path viable under c from f_{m-1} to $\mathcal{O}(f_m)$. Now, recall that for any f, and any cube $c^* \supseteq U$, we have:

$$(\mathcal{S}_U f)_{c^*} = f_{c^* - U}.$$

Using this identity, we can consider the macroexpanded subnetwork for f_m. Since the inputs for f_m not in U may be taken as specified by some cube c^*, we wish to consider the network $\mathcal{T}(f_m)_{c^*}$. Since the function $\frac{\partial \mathcal{O}(f_m)}{\partial f_{m-1}}$ has a satisfying assignment, a test exists for both stuck-at-0 and stuck-at-1 on the input lead $\mathcal{O}(f_{m-1})$ for this network. Now we must show:

Lemma 2.5.1 *A shortest path through any node f in a non-trivial network is viable under the cube 1, and hence under any cube c.*

Proof: Induction on $\delta(f)$. If f is a primary input, trivial. Assume for all f s.t. $\delta(f) < L$. If $\delta(f) = L$, let $\{\{P, h\}, f\}$ be a shortest path through f. By induction, the path $\{P, h\}$ is viable under 1 and there is a path viable under 1 through each input of f. Since each such path must be at least as long as $\{P, h\}$, each side input of f is late under every cube, and so under 1. Now, it is trivial that for any non-zero function g, $\mathcal{S}_{fanins(g)}g = 1$, and hence when U is equal to the entire set of side inputs of f we have that $\mathcal{S}_U \frac{\partial f}{\partial h} = 1$, thus $\{P, h, f\}$ is viable under 1. ■

Theorem 2.5.1 *Let N be any network, $\mathcal{T}(N)$ be the network obtained by any transformation $\mathcal{T}$ satisfying definition 2.5.1. Then for every viable path P in N terminating in f_m, there is at least one viable path $\mathcal{T}(P)$ in $\mathcal{T}(N)$ terminating in $\mathcal{O}(f_m)$, with $d(P) = d(\mathcal{T}(P))$.*

Proof: Consider some viable path $P = \{f_0, ..., f_m\}$ in N, its viability cube c and its transformation, $\{\mathcal{T}(f_0), ..., \mathcal{T}(f_m)\}$. Induction on $\delta(f_m)$. For $\delta(f_m) = 0$, the result is trivial, since f_0 is then a primary input. Assume for a path with $\delta(f_m) < L$. Suppose a path with $\delta(f_m) = L$. Then by assumption there is a path κ in $\mathcal{T}(N)$, viable under c, of length $d(\{f_0, ..., f_{m-1}\})$ terminating in $\mathcal{O}(f_{m-1})$. All that must be shown is the existence of an extension ν of κ of length $w(f_m)$, viable under c, from $\mathcal{O}(f_{m-1})$ to $\mathcal{O}(f_m)$. Since $\{f_0, ..., f_m\}$ is viable under c, there is a set of late inputs U under c such that $\mathcal{S}_U \frac{\partial f_m}{\partial f_{m-1}}$ is satisfied by c. By induction, each of the viable paths to these side inputs in the original network produces a viable path within the transformed network, so the same set U of inputs may be chosen to be late in $\mathcal{T}(N)$. The remaining inputs to $\mathcal{T}(f_m)$ have delays smaller than τ_{i-1} in the initial network, and so we must take their values as their static values under c. These values form a cube, c^*. This cube may be taken to propagate through the network $\mathcal{T}(f_m)$ producing $\mathcal{T}(f_m)_{c^*}$, and hence there is a viable extension of κ from $\mathcal{O}(f_{m-1})$ to $\mathcal{O}(f_m)$ in $\mathcal{T}(f_m)$ iff there is a viable path from $\mathcal{O}(f_{m-1})$ to $\mathcal{O}(f_m)$ in the cofactored network $(\mathcal{T}(f_m))_{c^*}$. Each path through the $(\mathcal{T}(f_m))_{c^*}$ has its length given by the arrival time of the input at its head. Since the only inputs to the cofactored network are $\mathcal{O}(f_{m-1})$ and the late side inputs, if any path from $\mathcal{O}(f_{m-1})$ to $\mathcal{O}(f_m)$ exists it is a shortest path. Hence by lemma 2.5.1 if such a path exists it is viable. We know that such a path exists since $\mathcal{S}_U \frac{\partial f_m}{\partial f_{m-1}}$ is satisfied by c, and so $f_{m_{c^*}}$ is a non-trivial function of f_{m-1}. Let ν be this path. By construction, ν is of length $w(f_m)$. By construction, the delay to $\mathcal{O}(f_m)$ through κ and the prefix of ν is τ_{m-1}, and $\{\kappa, \nu\}$ is viable under c in $\mathcal{T}(N)$. ■

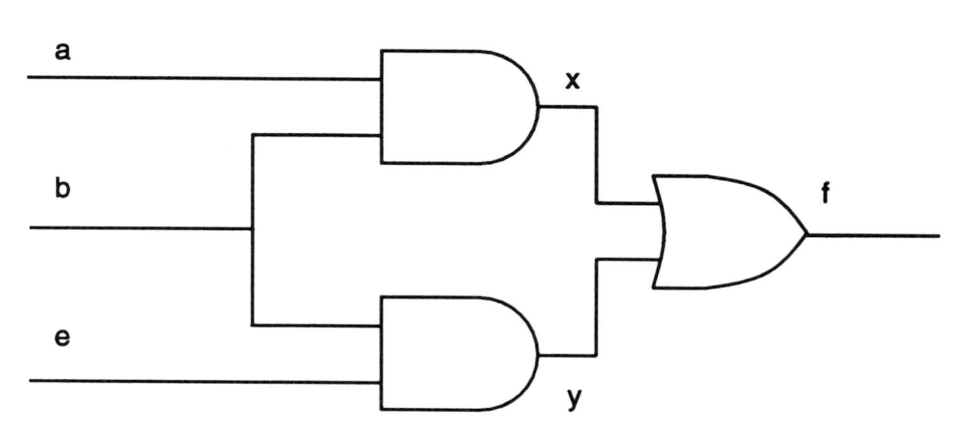

Figure 2.7: And/Or Transform of $f = ab + eb$

The converse to this theorem is false. Consider the circuit in figure 2.7. This is the and/or transform circuit of $f = ab + eb$. Now, suppose the variable b is late, and the value of e is set to 1. The function of the original gate is now $ab + b$, which simplifies to b: the reader can easily verify that $\mathcal{S}_b \frac{\partial f}{\partial a} = \overline{e}$, and so this gate is not sensitized to a when $e = 1$. However, in the transformed network, the path $\{a, x, f\}$ is viable when $e = 1$, since the variable y at the or-gate is late.

The fact that the converse is false has an important consequence: we cannot conclude that monotone speedup holds for arbitrary networks. However, we are guaranteed correctness of our algorithms if we transform the network N into a symmetric one, since theorem 2.5.1 guarantees that for each viable path in the original network we will find a viable path of equal length in the transformed network.

In the next chapter, we shall turn to algorithms. The general method is quite clear; transform the network into the and/or network, and then find the longest viable path in this network. We first introduce a mathematical tool that permits us to view viable paths as the satisfying set of a logic function; this in turn permits the development of a dynamic programming procedure to compute viable paths.

2.6 The Viability Function

At some very fundamental level, any set condition may be expressed as a multiple input logic function. We are interested in making explicit the logic function that underlies viability, because the computation of the viable paths may be made more efficient through explicit computation of this function, because various properties may be proved through use of this function, and because we shall develop a powerful theorem which permits us to quickly establish the correctness of approximation procedures.

The viability function ψ_P on a path P is easy to define: $\psi_P(c) = 1$ iff P is viable under c. This hardly gives more insight than the viable path definition. We prefer to define ψ_P in terms of some function $\psi_P^{f_i}$ at each node f_i. We develop this function intuitively, then justify its definition in a theorem at the end of this section.

Intuitively, we expect that the function ψ_P will be the product of the viability conditions at each node f_i along the path P, which in turn are captured in the function $\psi_P^{f_i}$

$$\psi_P = \prod_{i=0}^{m} \psi_P^{f_i}$$

Since time plays an important part in the definition of viable path, it is convenient to capture it in the definition of the viability function. For any node g, let $\mathcal{P}_{g,t}$ be the set of paths terminating in g of length at least t. Define

$$\psi^{g,t} = \sum_{Q \in \mathcal{P}_{g,t}} \psi_Q$$

Letting U be any subset of $S(f_i, P)$ we can express the viability condition on the subset as:

$$(\mathcal{S}_U \frac{\partial f_i}{\partial f_{i-1}}) \prod_{g \in U} \psi^{g,\tau_i - 1}$$

since this condition must be satisfied for one subset, we may write:

$$\psi_P^{f_i} = \sum_{U \subseteq S(f_i,P)} (\mathcal{S}_U \frac{\partial f_i}{\partial f_{i-1}}) \prod_{g \in U} \psi^{g,\tau_i - 1}$$

In summary, we define:

Definition 2.6.1 *The set of paths which terminate in g of length $\geq t$ are denoted $\mathcal{P}_{g,t}$.*

Definition 2.6.2 *The* **viability function** *(also* **viable set***) of a path $P = \{f_0, ..., f_m\}$ is defined as:*

$$\psi_P = \prod_{i=0}^{m} \psi_P^{f_i} \tag{2.2}$$

where

$$\psi_P^{f_i} = \sum_{U \subseteq S(f_i,P)} (\mathcal{S}_U \frac{\partial f_i}{\partial f_{i-1}}) \prod_{g \in U} \psi^{g,\tau_{i-1}} \tag{2.3}$$

and

$$\psi^{g,t} = \sum_{Q \in \mathcal{P}_{g,t}} \psi_Q \tag{2.4}$$

We have immediately:

Theorem 2.6.1 *$P = \{f_0, ..., f_m\}$ is viable under some minterm c iff c satisfies ψ_P.*

Proof:
$\Longrightarrow$ $P = \{f_0, ..., f_m\}$ is viable under c. Induction on $\delta(f_m)$. The base case is trivial, so assume for all paths with $\delta(f_m) < L$. Let $\delta(f_m) = L$. We must show that for each f_i, $c \in \psi_P^{f_i}$. Now, if $c \in \frac{\partial f_i}{\partial f_{i-1}}$, done. Otherwise, since the path is viable under c, we must have that there is a subset $U = \{g_1, ..., g_k\}$ of $S(f_i, P)$, where each g_j terminates a path Q_j which is viable under c and $d(Q_j) \geq \tau_{i-1}$. Since $\delta(g_j) < L$, by induction then $c \in \psi_{Q_j}$ whence $c \in \psi^{g_j,\tau_{i-1}}$ for every j. Further, $c \in \mathcal{S}_U \frac{\partial f_i}{\partial f_{i-1}}$, and done.
$\Longleftarrow$ $c \in \psi_P$. Induction on $\delta(f_m)$. The base case is trivial, so assume for all paths with $\delta(f_m) < L$. Let $\delta(f_m) = L$. We must show that the definition of viability holds at each node f_i on the path. Now, if $c \in \frac{\partial f_i}{\partial f_{i-1}}$, done. Otherwise, we must show that there exists a set of side inputs U meeting the conditions of the definition of viable path with $S_U \frac{\partial f_i}{\partial f_{i-1}} \supseteq c$. Since $c \in \psi_P^{f_i}$, then we must have that there is a subset $U = \{g_1, ..., g_k\}$ of the side inputs, and for each g_j, $c \in \psi^{g_j,\tau_{i-1}}$. Now, by the definition of $\psi^{g_j,\tau_{i-1}}$, we must have

$$c \in \sum_{Q_{jl} \in \mathcal{P}_{g_j,\tau_{i-1}}} \psi_{Q_{jl}}$$

Since c is a minterm, for each j it must be in $\psi_{Q_{jl}}$ for some k, and since $\delta(g_j) < L$, by induction Q_{jl} is viable under c, and $c \in \mathcal{S}_U \frac{\partial f_i}{\partial f_{i-1}}$, and so done. ∎

Observe that $\psi_P^{f_i}$ is a series, with one term for each subset U of the side inputs. The flaw in using static sensitization is that only one term of this series is taken (the term for $U = \emptyset$).

2.7 Summary

In this chapter, we've demonstrated that the two most obvious criteria for determining whether or not a path in a general logic network is false are either wrong or non-robust, and we've given a criterion that we can demonstrate is both correct and robust. Before we leave this chapter, let's revisit the criterion and examine its meaning intuitively.

The criterion consists of three equations, (2.2)-(2.4). Of these, all (2.2) says is that a path is true if and only if (2.3) is satisfied at every i. (2.4) merely captures the set of conditions under which a path terminating in g of length $\geq t$ is viable. The equation we are left to analyze for intuition, then, is (2.3).

$$\psi_P^{f_i} = \sum_{U \subseteq S(f_i,P)} (\mathcal{S}_U \frac{\partial f_i}{\partial f_{i-1}}) \prod_{g \in U} \psi^{g,\tau_{i-1}}$$

Using operator multiplication ($\mathcal{S}_{xy} f = \mathcal{S}_x \mathcal{S}_y f$), we can rewrite this equation as follows:

$$\psi_P^{f_i} = \left[\prod_{g \in S(f_i,P)} (1 + \mathcal{S}_g \psi^{g,\tau_{i-1}}) \right] \frac{\partial f_i}{\partial f_{i-1}}$$

Now, for purposes of intuition, suppose f_i is a simple gate. $\frac{\partial f_i}{\partial f_{i-1}}$ asserts a noncontrolling value on side input g. Let's examine the effect of the clause $(1 + \mathcal{S}_g \psi^{g,\tau_{i-1}})$, applied to $\frac{\partial f_i}{\partial f_{i-1}}$. Clearly multiplying $\frac{\partial f_i}{\partial f_{i-1}}$ by 1 merely yields $\frac{\partial f_i}{\partial f_{i-1}}$. However, the effect of multiplying $\frac{\partial f_i}{\partial f_{i-1}}$ by $\mathcal{S}_g \psi^{g,\tau_{i-1}}$) is interesting. A little manipulation shows that this is equal to $(\mathcal{S}_g \frac{\partial f_i}{\partial f_{i-1}})\psi^{g,\tau_{i-1}}$) which is the function $\frac{\partial f_i}{\partial f_{i-1}}$ with the assertion of a non-controlling value on g replaced by the function "g is late". Hence the viability criterion can be stated, colloquially, *for each side input g, either g is set to a noncontrolling value by primary input vector c, or g is delayed by input vector c*; in other words, c forces g's value to be either statically non-controlling or undetermined at $t = \tau_{i-1}$. The rationale behind the first is that if g's static value is non-controlling, it is certainly possible for g's final value to have arrived at $t = \tau_{i-1}$. The rationale behind the second is that if we do not know g's value at a precise instant in time (and, in general, we do not know g's value between the time of its first and last event), then we must assume that it is at a non-controlling value.

Chapter 3

False Path Detection Algorithms

Once a correct, robust sensitization criterion has been found, there remains the task of incorporating this criterion in an algorithm which finds the longest path satisfying this criterion; such a path is often called a *longest true path.* The development of such an algorithm is the subject of this chapter and the argument that is to be made is twofold. First, the methods that have appeared in the literature thus far which claim to solve this problem may be viewed as different parameterizations of a single algorithm, and, second, that this algorithm can be modified to compute the viability procedure corresponding to the viability criterion devised in the last chapter.

All of the algorithms that have appeared in the literature to date are of one broad, generic, parameterized class. A collection of partial paths, each of which is known to be true, is maintained. At each step of the algorithm, one such partial path, say $P = \{f_0, ..., f_m\}$ is removed from the structure. If f_m is an output of the circuit, then this is a full true path, the fact is recorded and (if this is a so-called *best-first procedure*) the procedure terminates. If this is not an output, then some unexamined fanout of f_m, say g, is selected to extend the path. A boolean function $\gamma(P, g)$ is computed. If $\gamma(P, g)$ is satisfied, then P, g is a true path and is inserted into the structure of true paths. This procedure continues until a termination criterion is met.

The unity of the algorithms which address this question is not generally recognized; in particular, the identification of the sensitization criterion with a logic function is not usually made. Procedures based on the D-Algorithm (e.g., [11] [6]) compute this function *implicitly*. Nevertheless, it is important to recognize that such a function exists for each algorithm, and indeed, in the case of the papers cited, has a simple explicit form, which we shall divulge in the sequel.

The parameters to this generic procedure are:

1. The search method, which is expressed in terms of the maintained data structure of true paths and in the termination conditions;
2. The definition of what constitutes a true path, the so-called *sensitization conditions*, which expresses itself in the choice of the boolean function γ; and
3. The method used to determine whether the sensitization function is satisfiable.

The choice of these parameters are largely independent. The choice of search method affects the computational complexity of the procedure, while the choice of sensitization condition and satisfiability test affect the *tightness* and also the *correctness* of the procedure: a false-path procedure is *tight* if it provides a least upper bound on the delay of the longest true path; a false-path procedure is *incorrect* if it underestimates the delay before a circuit output settles to its final value.

3.1 Generic False Path Detection Algorithm

We begin our discussion of the single, generic, procedure by mentioning that most authors in the field would argue that there are at least two different methods, *depth-first* and *best-first* search[6]. To a large extent whether one considers these different parameterizations of a generic procedure or two different procedures is a matter of taste; one can, after all, view *any* algorithm as an appropriate parameterization of a Universal Turing Machine. We hope to show that the two search procedures can be viewed as the same basic routine, differing only in termination condition and in the data structure used to store the partial paths.

The *best-first* procedure maintains the partial true paths in a priority queue ordered by the *potential full length*, or length of the longest extension, of the partial path; this quantity is named the *esperance* of the partial path[6]. The best-first procedure terminates when an output is reached, since by construction no longer true path can exist. The depth-first procedure maintains the partial paths on a LIFO stack. When a full path is reached, the path is examined to see if it is of greater length than the longest full true path found so far; if it is, then this is recorded. When the stack is empty, the procedure terminates, and the best path found is returned.

The best-first procedure is slightly more complex than the depth-first procedure, and must be carefully implemented. If it is, then at most KD paths are examined by the procedure, where K is the number of long false paths and D the diameter of the graph. The depth-first procedure will in general examine an exponential number of paths.

3.1.1 Depth-First Search

Depth-first search is a classic graph search algorithm. The central idea is that the fanouts of every node in the graph are ordered, and the subgraph headed by each fanout of a node is explored in its entirety before the succeeding fanout in the order is examined.

The basic, recursive depth-first procedure is depicted in figure 3.1. The basic routine in this code is the function `find_path_dfs`. This function takes one argument, the current true partial path, and returns the longest true full path containing this partial path as a prefix.

The code is fairly self-explanatory. If the partial true path is a full path, then simply return. Otherwise, if there is a longest true full path containing this partial path as a prefix, then this path must be obtainable through some output of the last node of the path, named `node` in this code. Hence, for each such output, attempt to extend the current partial path; if such an extension succeeds, then the answer is to be found by calling the procedure recursively on the obtained successor partial path; finding the longest path over all fanouts yields the solution.

Of course, any recursive procedure can be phrased as an iterative procedure by keeping a last-in, first-out (LIFO) stack; the stack maintains the information that otherwise is implicitly maintained by the sequence of outstanding function calls (in fact, this is precisely the way a machine keeps track of the various function arguments to a procedure). A `push` onto this stack is equivalent to a recursive call; a `pop` to a return. For the basic depth-first search procedure, this rephrasing is well worth doing. First, exposing the underlying stack structure inherent in a recursion yields insight into the unity of this approach with the best-first

```
find_longest_true_path() {
    max_length ← 0;
    long_path ← ∅
    foreach primary input p {
        path ← find_path_dfs(p);
        length ← length(path);
        if(length > max_length) {
            max_length ← length;
            long_path ← path;
        }
    }
}
find_path_dfs(path) {
    node is the last node on path;
    if(output(node)) return path;
    else {
        max_len ← 0;
        long_path ← ∅;
        foreach fanout p of node {
            if(γ{p, path} ≢ 0) {
                long1 ← find_path_dfs({p, path});
                length1 ← length(long1);
                if(length1 > max_len) {
                    max_len ← length1;
                    long_path ← long1;
                }
            }
        }
    }
}
```

Figure 3.1: Recursive Depth-First False Path Detection Algorithm

approach; and, second, this permits the search to be easily pruned. We give the code for the iterative version of the depth-first procedure in figure 3.2.

The items which need to be stacked are the explicit argument to `find_path_dfs`, namely the current partial path, and some local variables of that routine. In this case, the only such variable is the fanout next to be explored (the variable `p`). We represent this as a counter (`path.next_fanout`) associated with the partial path.

An improvement is possible to this routine. In general, this procedure will explore an exponential (in the number of nodes of the graph) number of paths, and so take a very long time to perform the computation. In practice, the search space can be pruned. We are only interested in the longest true path, or in a set of such true paths. If the longest possible extension of a given partial path is shorter than the longest path already found, then there is no point in exploring any extension at all of this path, and one might as well terminate the search immediately. This is easily accomplished. The longest possible extension of a given partial path `path` which terminates in `node` is given by the conjunction of `path` with the longest path originating in `node`. The length of this path is deduced easily in linear time for all the nodes in the network, and may be stored at each node. The code for this calculation is given in figure 3.3.

Once the calculation of the longest path from every node is done, pruning the search space is easy. In figure 3.2, the line

```
if(γ({path, g})≢ 0)
```

is replaced by the line

```
if(((length(path) + best_path_from(g)) > max_length)
and (γ({path, g})≢ 0))
```

to obtain the variant of the algorithm with pruning.

Pruning is an effective heuristic technique. Despite this, however, the complexity of depth-first search is still exponential in the number of nodes; there is no guarantee that the pruning technique will eliminate a substantial fraction of the paths. It would be better if *only* the longest true path and the longer false paths are examined. Since these paths must be examined by any algorithm which purports to solve this problem, this procedure is quite efficient.

```
find_longest_true_path() {
    Initialize stack to primary inputs of the circuit
    max_length ← 0;
    long_path ← ∅;
    while(path ← top(stack) ≠ ∅) {
        k is the last node on path;
        if(k is an output) {
            if(length(path) > max_length) {
                max_length ← length(path);
                long_path ← path;
            }
        }
        if(path.next_fanout > k.num_fanouts) {
            pop path from stack;
        } else {
            g ← k.fanouts[path.next_fanout];
            path.next_fanout = path.next_fanout + 1;
            if(γ({path, g})≢ 0) {
                new_path ← {path, g} is true;
                push new_path on stack;
                new_path.next_fanout ← 0;
            }
        }
    }
    return long_path;
}
```

Figure 3.2: Depth-First False Path Detection Algorithm

```
longest_path_from_nodes() {
    nodes ← array sorted in topological order;
    for i ← |nodes| downto 0 {
        length ← 0;
        foreach fanout p of nodes[i] {
            if(length < best_path_from(p))
                length ← best_path_from(p);
        }
        best_path_from(nodes[i]) ← length + weight(nodes[i]);
    }
}
```

Figure 3.3: Procedure Calculating the Longest Path from Every Node

3.1.2 Best-First Search

The inefficiency in the depth-first procedure arises from the fact that the path removed from the stack at each iteration is the last path shoved on the stack; this makes both the **push** and **pop** operations $O(1)$, but gives a deleterious effect on the performance of the algorithm as a whole. It would be better if the partial path of maximum potential length were removed from the stack at every iteration. Indeed, if this were done one might expect a polynomial bound on the complexity of the algorithm if there were not an exponential number of long false paths.

This assurance can be given by revising the data structure underlying the iterative construction. If a *priority queue* is used instead of a LIFO stack, then this guarantee can in fact be given.

A priority queue is a data structure with two major properties.

1. At each iteration, the maximum element of the queue (with respect to some standard order) is at the head of the queue

2. A sequence of n *enqueue* and *dequeue* operations takes $O(n \log n)$ time.

Priority queues are typically implemented on top of *heaps*. A *heap* is defined as a full binary tree with the property that every element is greater than each of its descendants. A full discussion of priority queues and heaps can be found in any good sophomore or junior algorithms text; we took our implementation from [79].

The priority queue is ordered by the *esperance*[6] of a *minimal extension* of a partial path.

Definition 3.1.1 *An* **extension** *of a partial path $P = \{f_0, ..., f_n\}$ is a partial path Q such that P is a prefix of Q. A* **minimal extension** *of P is an extension $\{P, f\}$ where f is a single node. A* **full extension** *of a path P is any extension of P terminating in a primary output.*

Definition 3.1.2 *The* **esperance** *of a partial path P denoted $E(P)$, is defined as the length of the longest full extension Q of P.*

Definition 3.1.3 *The set of* **unexplored extensions** *of a path P with respect to a priority queue, denoted $UE(P)$, is the set of minimal extensions Q of P such that no extensions of Q appear on the queue.*

Note that the esperance of a partial path P is greater than or equal to the esperance of any extension of P.

The operation of the best-first procedure is that, at each iteration, the path with the longest full extension is popped off the queue and extended through the minimal extension with the greatest esperance. Since there is little point in extending a partial path through the same minimal extension twice, the list of minimal extensions through which a given partial path has not been extended is crucial; this is the set of unexplored minimal extensions. Similarly, the metric of interest for determining which path should be extended on the next iteration is not its longest full extension, but rather its longest full extension through an unexplored minimal extension (other full extensions have already been "covered" by preceding extensions). Hence the priority queue is ordered by the esperance of a path through an unexplored minimal extension.

Theorem 3.1.1 *For every partial path P, we have:*

$$E(P) = \begin{cases} d(P) & P \text{ terminates in an output} \\ d(P) + \max_{h_0 \in FO(P)} D(h_0) & \text{otherwise} \end{cases}$$

Proof: If P terminates in an output, then trivial. Otherwise, let Q be the longest full extension of P. Since every full extension of P is obtained through some minimal extension $\{P, h_0\}$ of P, and since $D(h_0)$ is the length of the longest path from h_0 to a primary output, we must have:

$$E(P) = d(P) + D(h_0)$$

for some fanout h_0 of P. ■

Note that this theorem implicitly requires that the fanouts of a node be explored in order of decreasing maximum distance from a primary output. Hence we assume that the fanouts of each node have been sorted in decreasing order by maximum distance from a primary output.

We show the code for the best-first procedure in figure 3.4.

We now prove that the algorithm of figure 3.4 is correct, i.e., returns the longest path true by the sensitization criterion γ. The technique used in the proof of this theorem is the classic *loop invariance* technique[34]. In some sense, this is the program correctness version of induction. In this proof technique, a statement is shown to hold on the $N+1$st iteration of the loop if it holds on the Nth, and is also shown to hold upon entry into the loop.

Theorem 3.1.2 *Through each loop of the algorithm every true path Q has a prefix P on the queue such that $E(P) \geq d(Q)$.*

Proof: Loop invariance. As the main loop of the algorithm is entered, this is clearly true, for the set of esperances of the partial paths consisting of only the inputs is equal to the lengths of the longest paths from each input, and one of these is clearly at least as long as the longest true path. Suppose true through N iterations. On the $N + 1$st iteration, if the condition is violated then let Q_1 be the true path. Since the condition held through N iterations, then $Q_1 = \{P_1, h_0, ..., h_n\}$ for some P_1, and P_1 was on the queue through the Nth iteration with $E(P_1) \geq d(Q)$. Further, on the $N + 1$st iteration either P_1 is no longer on the queue, or $E(P_1) < d(Q)$. The latter case can only occur if every extension of P_1 with esperance $\geq d(Q)$ has been rejected as false; but the extension $\{P_1, h_0\}$ is true and has esperance $\geq d(Q)$, and so this cannot occur. Similarly, if P_1 has been removed from the queue, then every extension of P_1 through one of its fanouts has been processed, and those found to be true inserted on the queue; this set includes $\{P_1, h_0\}$, and this has

```
find_longest_true_path() {
    Initialize queue to primary inputs of the circuit
    while((path ← pop(queue)) ≠ nil) {
        k is last node on path;
        if(k is an output) return path;
        if(path.next_fanout ≤ k.num_fanouts) {
            g ← k.fanouts[path.next_fanout];
            if(γ({path, g})≢ 0) {
                new_path ← {path, g} is true;
                insert new_path on queue;
                new_path.next_fanout ← 0;
            }
            path.next_fanout = path.next_fanout + 1;
            if(path.next_fanout ≤ k.num_fanouts) {
                E(path,path.next_fanout) ← d(path) +
                    best_path_from(k.fanouts[path.next_fanout]);
                insert path on queue;
            }
        }
    }
}
```

Figure 3.4: Best-First False Path Detection Algorithm

esperance $\geq d(Q)$; hence we conclude the statement holds through $N+1$ iterations ■

Corollary 3.1.3 `find_longest_true_path()` *finds a longest true path.*

Proof: Let Q be a longest true path, Q_1 the longest true path reported by the best-first procedure. All we must show is that $d(Q) = d(Q_1)$. By the theorem, Q has a prefix P on the queue through each loop with $E(P) \geq d(Q)$. But since Q_1 was at the top of the queue on the last iteration, we must have that $E(Q_1) \geq E(P) \geq d(Q)$, and, since $E(Q_1) = d(Q_1)$, we have that $d(Q_1) \geq d(Q)$, hence Q_1 is a longest true path. ■

For the efficiency of this algorithm, we note the following. Let Q be the longest true path reported by algorithm 3.4. If there are K full false paths longer than Q, and if the diameter of the graph is D, then at most KD partial paths have esperance greater than Q. (This upper bound is obtained through the observation that each full path has at most D prefixes, and the only paths with esperances greater than Q are the false paths and their prefixes). At each loop of the algorithm, one of these partial path or some prefix of Q was examined and extended or rejected as false. There are at most $D(K+1)$ such paths, and hence at most $D(K+1)$ iterations of the algorithm. For the general algorithm, the cost of each iteration is dominated by the determination of whether or not a new partial path is true, which we denote by S[1]; this cost is therefore

$$O(KDS)$$

The remainder of the cost of the algorithm is dominated by the insertions and deletions from the priority queue. The cost of a single insertion or deletion on a priority queue containing n elements is well-known to be $O(\log n)$. There are at most $O(D)$ elements on the queue at any time, and hence for $O(KD)$ insertions and deletions we have a cost of:

$$O(KD \log D)$$

and hence the cost of the algorithm is:

$$O(KD \log D + KDS)$$

The depth-first and best-first procedures have been compared [6] [92]. The former experiments concluded that the depth-first procedure with pruning outperformed the best-first procedure by a wide margin; this result is surprising and anomalous, given that a complexity analysis would indicate that the overhead of the best-first procedure is at most logarithmic, while the overhead of the depth-first procedure is in general exponential.

The latter set of experiments [92] concluded that the best-first procedure outperformed the depth-first procedure in the related problem of finding the n longest paths in a directed acyclic graph. This result one would expect; however, their experiments also concluded that the margin was very slight, much less than one would expect.

[1] As we see in appendix A, this problem is $\mathcal{NP}$-complete for most definitions of γ

The variance in the best- vs depth-first procedures reported in [6] offers one possible explanation. A minor error in the implementation of the best-first procedure can lead to a large number of paths being searched. If, when a path P is found true, *every* one-node extension of P (as opposed to merely the best) is placed on the queue, then if every node has an average fanout of k, tracing a single path to the output will result in kD nodes being placed on the queue. Note in the best-first procedure given above only *one* extension of P is explored when P is found true, and hence only D paths are placed on the queue as P is explored.

3.1.3 Generic Procedure

Now, notice that the code in algorithms 3.4 and 3.2 are very similar; the principal distinction is in the data structure used to represent the set of active partial paths. A second difference is in the termination condition used and in the steps the algorithm performs when it finds a full true path. The distinction between priority queue and stack has been treated adequately above. The difference in termination condition and action on finding a full true path is an artifact of the fact that the best-first procedure is so constructed that first true path found is also a longest true path; the depth-first procedure offers no such assurance, so the search must be continued, after recording the fact that a new long path has been found.

Nevertheless, the two procedures are similar enough that one can consider them a single generic procedure, parameterized by search method. The code is shown in figure 3.5.

The generic procedure can easily be modified to permit pruning under depth-first search.

3.1.4 Modifying the Generic Procedure to find Every Long True Path

In some applications (for example, resynthesis for timing) it is desirable not only to find the longest true path, but also every true path within some ϵ of the longest true path; the rationale is that if the longest true path fails to meet timing specifications, and the circuit must be resynthesized for timing [82], then it is of little use to modify only the critical path if another series of paths remain true and equally long. Another

```
find_longest_true_path() {
    Initialize paths to primary inputs of the circuit
    if(depth_first_search)
        long_path ← ∅; maxlen ← 0;
    while(path ← pop(paths)) {
        k is last node on path;
        if(k is an output) {
            if(depth_first_search) {
                if (d(path) > maxlen) {
                    maxlen = d(path);
                    long_path ← path;
                }
            }
            else return path;
        }
        if(path.next_fanout ≤ k.num_fanouts) {
            g ← k.fanouts[path.next_fanout];
            path.next_fanout = path.next_fanout + 1;
            if(path.next_fanout ≤ k.num_fanouts) {
                E(path) ← d(path) +
                    best_path_from(k.fanouts[path.next_fanout]);
                insert path on paths;
            }
            if(γ({path, g}) ≢ 0) {
                new_path ← {path, g} is true;
                insert new_path on paths;
                new_path.next_fanout ← 0;
            }
        }
    }
    if(depth_first_search) return long_path;
    else return ∅;
}
```

Figure 3.5: Generic False Path Detection Algorithm

variant on this procedure is to report every true path of length greater than some threshold, which represents the maximum allowable delay of the circuitry.

These two problems require the same modification to the basic algorithm; a threshold T is computed, and all paths of length $> T$ are reported; in the case of the first variant, $T = L - \epsilon$, where L is the length of the longest path and ϵ is the user-supplied tolerance. In the case of the second variant, T is user-supplied.

The modification to the generic procedure is conceptually fairly simple; the routine merely maintains a list of paths of length $> T$, and returns this list once it is clear that no further paths will be found of length $> T$; in the case of the best-first procedure, this occurs when the top path on the queue has *esperance* $\leq T$; in the case of the depth-first procedure, this is simply when the stack is empty, though pruning can be used effectively here, as well.

A little bookkeeping is required when the threshold is set dynamically to $L - \epsilon$. Again, in the case of the best-first procedure, this is a fairly simple matter; the threshold is initially set to 0, and when the longest path is found the threshold is set to $L - \epsilon$. In the case of a depth-first search, the threshold is recalculated every time a new longest path is found. The code is shown in figure 3.6

One interesting note here is that the time of the best-first procedure to find all paths of length greater than the threshold is determined simply by the number of paths, false and true, of length greater than the threshold. This may or may not be greater than the total number of long false paths in the circuit, if the threshold is fixed.

3.1.5 Varying Input Times, Output Times, and Slacks

In practice, system requirements often dictate that inputs to a circuit arrive at differing times, or that outputs are required at varying times, or both. Timing analyzers often take this into account by computing the arrival and required times for a signal separately (using the analogous formula for required time), and then compute the *slack* for each node as the difference between the required and arrival time; it is easy to show that there is at least one sequence of nodes with the minimum slack, and this is defined as a *critical path* of the circuit.

We define the *slack* of a path P, $slack(P)$ as follows. Let $P = \{i_k, f_0, ..., f_m, o_j\}$, input arrival time of i_k is t_k, output required time of

```
find_longest_true_path() {
    Initialize paths to primary inputs of the circuit
    long_paths ← ∅;
    if(depth_first_search) maxlen ← 0;
    if(search_by_epsilon) T ← 0;
    while(path ← pop(paths)) {
        k is last node on path;
        if(k is an output)
            output_reached(path, T, long_paths, ε, maxlen);
        if(best_first_search and E(path) ≤ T) return long_paths;
        if(path.next_fanout ≤ k.num_fanouts) {
            g ← k.fanouts[path.next_fanout];
            path.next_fanout = path.next_fanout + 1;
            if(path.next_fanout ≤ k.num_fanouts) {
                E(path) ← d(path) +
                    best_path_from(k.fanouts[path.next_fanout]);
                insert path on paths;
            }
            if(γ({path, g}) ≢ 0) {
                new_path ← {path, g} is true; new_path.next_fanout ← 0;
                insert new_path on paths;
            }
        }
    }
    return long_paths;
}
```

Figure 3.6: Procedure Returning All Longest Paths

```
output_reached(path, T, long_paths, ε, maxlen) {
    if(d(path) > T) long_paths ← long_paths ∪ {path};
    if(depth_first_search and d(path) > maxlen and search_by_epsilon) {
        must update the threshold and determine which paths still
        are longer than the threshold
        T ← d(path) - ε; new_long_paths ← {path};
        foreach path p1 on long_paths {
            if (d(p1) > T) {
                new_long_paths ← {p1} ∪ new_long_paths;
                maxlen ← d(path); long_paths ← new_long_paths;
            }
        }
    }
    else if(T = 0) {
        long_paths ← {path};
        T ← d(path) - ε;
    }
}
```

Figure 3.7: Auxiliary Procedure to Procedure Returning All Longest Paths

o_j is t_j. Then the slack of P is defined:

$$slack(P) = t_j - [t_k + \sum_{i=0}^{m} w(f_i)].$$

Intuitively, the slack of P is the difference between the required time of o_j and its arrival time down the path. The *critical path* of a circuit is the true path of minimum slack.

It is possible to perform an analogous calculation and eliminate false paths, but this would require a significant restructuring of the algorithm given above. All things being equal, it would be better to adapt the existing algorithm by appropriate manipulation of the graph structure of the circuit.

Fortunately this adaptation is trivial. If inputs arrive at different times, we might as well consider the time when the earliest input i_1 arrives to be 0 (to avoid the inconvenience of negative weights). Let each input i_j arrive at $t = t_j$; we can then consider i_j as an internal node of the circuit, the output of a static delay buffer of weight t_j whose input is a new input i'_j, which arrives at $t = 0$. The resulting circuit has the property that every path originating in i_j has a corresponding path in the original circuit originating in i'_j of identical delay.

Similarly, if outputs have differing required times, let t_{max} be the maximum required time of all the outputs. To each output o_j with required time t_j attach a static delay buffer of weight $t_{max} - t_j$, input o_j, output o'_j. Hence the required time of o_j in the resulting circuit is t_j. The resulting circuit has the property that every path in the original circuit terminating in o_j with delay d has a corresponding path in the resulting circuit terminating in o'_j of delay $d + t_{max} - t_j$. Ideally, we wish to show that this transformation of the circuit graph preserves the set of critical paths of the circuit; in particular, we wish to show the following:

Theorem 3.1.4 *Let $P = \{i_k, f_0, ..., f_m, o_j\}$ be any path in the original circuit. Then $P' = \{i'_k, f_0, ..., f_m, o'_j\}$ is the corresponding path in the resulting circuit, and $slack(P) = slack(P')$.*

Proof:

$$\begin{aligned} slack(P) &= t_j - [t_k + \sum_{i=0}^{m} w(f_i)] \\ &= t_{max} - (t_{max} - t_j) - [t_k + \sum_{i=0}^{m} w(f_i)] \\ &= t_{max} - [t_k + (\sum_{i=0}^{m} w(f_i)) + (t_{max} - t_j)] \\ &= slack(P') \end{aligned}$$

■

Now, note that P' is true iff P is true, and, hence, if P is a critical path then P' is a critical path. Further the primary inputs in the resulting network all have arrival times of 0, and the primary outputs all have required times of t_{max}. Hence the critical path in the transformed circuit is the longest true path, as desired.

3.1.6 Separate Rise and Fall Delays

The models discussed so far assume that the delay in the response of a gate to one of its inputs is characterized by a single number, its *delay*. In fact, a common feature of modern logic circuits is that the delay in the response of a gate to an input is different, depending upon whether the input is undergoing a rising or falling transition. It is reasonable to ask whether the generic false path algorithm and its variants can handle separate rise and fall delays.

One addition is immediate: we must maintain in the path data structure the edge on the last node of the path, either rising (RISE), or falling (FALL). Below, we shall see how to maintain and compute the new edges.

If we assume for a moment that each gate in the circuit is monotonic in each input, then each path in the graph in fact represents two logical paths through the circuit: one path with a falling transition on the input at the head of the path, and one path with a rising transition on the input at the head of the path. Both logical paths may be represented by modifying the initialization of the data structure of partial paths. Rather than pushing a single path per primary input on the structure, we push two, one with a rising an done with a falling edge. Since the path beginning with a rising transition on primary input p can only be activated when the final value of p input is high (1), its initial sensitization function is not $\gamma = 1$ but rather $\gamma = p$; similary, the initial sensitization function for the path beginning with a falling transition on primary input p is $\gamma = \bar{p}$. In other words, the initialization sequence changes from:

foreach primary input p of the circuit
    push(path({p}, $\gamma = 1$, delay = 0)) on circuit

To:

foreach primary input p of the circuit
    push(path({p}, edge = RISE, $\gamma = p$, delay = 0)) on circuit
    push(path({p}, edge = FALL, $\gamma = \bar{p}$, delay = 0)) on circuit

Edges are updated according to the following table:

Input Edge	Gate inverting in input	Gate noninverting in input
RISE	FALL	RISE
FALL	RISE	FALL

Gates that are both increasing and decreasing in some input are handled by modelling each such gate as two gates, one increasing and one decreasing.

3.1.7 Don't-Care Conditions

Not all input conditions actually occur during operation of a circuit, and not all output values are used for all of the input conditions which do occur. It is important in timing analysis to ensure that the input vector which sensitizes a path actually occurs, and, further, that the output at the tail of a path is used when that input vector is applied to the primary inputs.

Input vectors which do not occur are collectively referred to as the *input don't-care* set; input vectors for which a particular output is unused is called that output's *output don't-care set.* Our constraint is then that the sensitization vector for a path terminating in primary output o_j lie in the complement of the *input don't-care set* and in the complement of the output don't-care set of o_j.

Ensuring that the sensitization vector lies outside the input don't-care set is achieved on initialization, merely by multiplying the sensitization function γ of each initial path by the complement of the input don't-care set. Similary, ensuring that the sensitization vector for a path terminating in primary output o_j lie in the complement of the output don't-care set of o_j is achieved simply by creating a formal primary output node for each primary output and setting the local sensitization function for that node to be the complement of the output don't-care set for that primary output.

3.2 Viability Analysis Procedure

The assumption underlying the generic procedure was that the function γ was a function only of the path being extended. An examination of the viability equations (2.2)-(2.4) and the surrounding discussion demonstrates that this assumption is unfulfilled by the viability function. The viability function of a path is not only a function of the path itself, but also of all adjoining paths at least as long; this dependence expresses itself in the subfunction ψ^{g,τ_i-1}. Correct computation of the viability function requires that this function be computed for each side input to each node on the candidate path as that node is encountered. Conceptually, this could be done by recursively tracing the set of viable paths terminating in each side input as a node is encountered on the candidate path.

3.2.1 Naive Depth-First Search Procedure

Our first algorithm is designed to find a viable path of length at least L, terminating in some target node f_i, with a prefix on a given `stack`. Conceptually, it is based on the iterative version of the depth-first procedure explored earlier in this chapter. It is shown in figure 3.8.

```
find_viable_path(stack, network, target_node, length)
{
    while((path ← pop(stack)) ≠ ∅) {
        if(last_node(path) ≡ target_node) return(path, stack);
        foreach fanout c of last_node(path) {
            if (path_length(path) + longest_path(c, target_node) < length)
                continue;
            if (c ≡ target_node) psi ← 1;
            else psi ← viable_set(c, last_node(path), network,path_length(path));
            new_psi ← psi * viability_function(path);
            if(new_psi ≢ 0) {
                new_path ← c,path;
                path_length(new_path) ← weight(c) + path_length(path);
                viability_function(new_path) ← new_psi;
                push(new_path, stack);
            }
        }
    }
    return (∅, ∅);
}
```

Figure 3.8: Naive Algorithm to Find the Longest Viable Path

If we assume for the moment that the **viable_set** computes the function:

$$\sum_{U \subseteq S(f_i,P)} (\mathcal{S}_U \frac{\partial f_i}{\partial f_{i-1}}) \prod_{g \in U} \psi^{g,\tau_i-1}$$

then the correctness of this algorithm is easily established. A partial path is popped off the stack; if it is complete, then we are done and return. Otherwise, each successor node is examined to see if it may extend this path fruitfully; if it can, the extended path is pushed on the stack. Those nodes which cannot possibly extend this path to the desired length at the target node, and those which are not viable, cannot extend this path. This process continues until either no paths may be extended (the stack is empty), or one path is complete. Note that in addition to returning a viable path of the appropriate length terminating in **target_node**, this routine also returns the final **stack**; this is to permit this routine to be called iteratively by a procedure which finds all the viable paths of the appropriate length.

It is now time to write the function which computes all the viable paths of length at least L, and which terminate in node **target_node**. It is shown in figure 3.9. Note this procedure consists simply of calling **find_viable_path** repeatedly until the **stack** is exhausted. The mechanism of passing the **stack** into and out of **find_viable_path** is simply a means of preserving the **stack** over calls to **find_viable_path**.

The correctness of this routine is easy to establish, given the correctness of **find_viable_path**, simply by observing that the correct set of initial paths are the paths consisting of only the primary inputs. These paths have viability function 1, and length 0.

With these procedures in hand, **viable_set** falls out easily from the definition above, and is shown in figure 3.10. This procedure simply finds all the viable paths of length τ_{i-1} terminating in each side input **k** to **c**, and sums up their viability functions in the field **k.psi**. The function $\psi^{\mathbf{k},\tau_{i-1}}$ is thus computed and stored in **k.psi**, and the viable set falls out easily by equation 2.3.

The algorithm to find the longest viable path is similarly easy (assuming we have made the obvious trivial change to **find_viable_path** so that the target node may be any one of a set). It is shown in figure 3.11

This, however, may cause the same partial path to be traced potentially many times, each time it is encountered as an abutting path to

```
find_all_viable_paths(node, network, length)
{
    list ← [];
    stack ← [];
    foreach primary_input p of network {
        P ← a new path of node p, length 0, psi ← 1;
        push P on stack;
    }
    /* Initialize path and stack */
    (path, stack) ← find_viable_path(stack, network, node, length);
    while(path ≠ ∅) {
        list ← list, path;
        (path, stack) ← find_viable_path(stack, network, node, length);
    }
    return list;
}
```

Figure 3.9: Naive Algorithm to Find All the Long Viable Paths

a shorter candidate path. A second alternative is to store the viability functions for each traced path, and only trace side paths recursively when these are known not to have been traced. A still better alternative would be to avoid the recursive path tracing at all. This can be done if:

1. It is known that every longer side path to the candidate path has been traced; and
2. The function $\psi^{g,\tau_i - 1}$ is maintained in a variable attached to g and is known to be correct, i.e., is known to contain the sum of the viability functions of all such longer side paths.

3.2.2 Dynamic Programming Procedure

These assurances can be given by a *dynamic programming* procedure based on the best-first procedure. Recall that the best-first procedure

```
viable_set(c, prev_node, network, length)
{
    psi ← 0;
    foreach input k of c, k ≠ prev_node {
        k.psi ← 0;
        list ← find_all_viable_paths(k, network, length);
        foreach path p on list
            k.psi = k.psi + path_psi(p);
    }
    foreach subset U of the side inputs{
        new_psi ← S_U ∂c/∂prev_node;
        foreach k in U
            new_psi ← new_psi * psi[k];
        psi ← psi + new_psi;
    }
    return psi;
}
```

Figure 3.10: Viable Set Algorithm

examines partial paths in decreasing order of their *esperance*, which is also their *potential full length.* As a result, the best-first procedure is effectively a longest-path-first procedure.

Simply using an ordering on esperance is insufficient to our purposes, however. If we are examining a node f_i and attempting to extend a path P of length τ_{i-1}, we must ensure that *all* side viable paths of P at f_i of length $\geq \tau_{i-1}$ have been examined. However, a side path of length precisely τ_{i-1} may well have esperance equal to $E(P)$. We must break this tie in favour of the path whose last node is of *lesser* level. With this in mind, we can define a successor relation $\succ$ on partial paths.

Definition 3.2.1 *Let* $Q = \{g_0, ..., g_n\}$, $P = \{f_0, ..., f_m\}$. $Q \succ P$ *iff* $E(Q) > E(P)$ *or* $E(P) = E(Q)$ *and* $\delta(g_n) < \delta(f_m)$.

```
find_longest_viable_path(network)
{
    foreach primary_input p of network {
        P ← a new path with nodelist ← {p}, length ← 0, psi ← 1;
        push P on stack;
    }
    (path, stack) ←
        find_viable_path(stack, network, primary_outputs(network), 0);
    while(path ≠ ∅) {
        oldpath ← path;
        length ← path_length(path);
        (path, stack) ←
            find_viable_path(stack, network, primary_outputs(network), length);
    }
    return oldpath;
}
```

Figure 3.11: Naive Algorithm to Find the Longest Viable Path

Lemma 3.2.1 *Let $P = \{f_0, ..., f_i\}$ be a partial path. Let $Q = \{g_0, ..., g_n\}$ be a side path to P at f_i. If $d(Q) \geq \tau_{i-1}$ then $Q \succ P$.*

Proof: Let P, Q be as stated in the premise of the lemma, $d(Q) \geq \tau_{i-1}$. Then $E(P) = \tau_{i-1} + w(f_i) + K$, where K is the maximum distance from P to a primary output. But $E(Q) \geq d(Q) + w(f_i) + K$, whence $E(P) \leq E(Q)$. Since $\delta(f_i) > \delta(g_n)$, $Q \succ P$. ∎

This result gives us the tool we need to avoid excessive computation of the viable sets if we replace the stack in `find_viable_path` with a priority queue of extensions ordered in descending order under $\succ$; thus, at each iteration, we attempt the extension that is maximal wrt $\succ$, and we are guaranteed that we have examined all partial paths Q such that $Q \succ P$. Note that this procedure is very similar to the best-first procedure, differing only marginally in the function by which the priority queue is ordered.

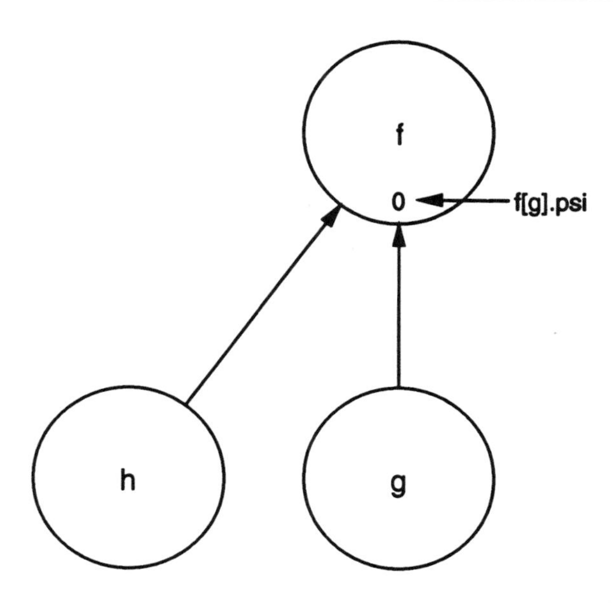

Figure 3.12: Viability Algorithm Variables

We need variables in which to keep the sums of the viability functions of the paths traced thus far. At each connection, from a node g to some node f we keep a field $f[g].psi$. Each such field is initially 0. As we pop an extension $\{P, f\}$ off the queue, ending in node g, we set $f[g].psi = f[g].psi + \psi_P$ The rationale is that the path P is certainly true, and since it was on the top of the queue it is certainly longer than any other path on the queue that may be extended through f. A depiction of the dynamic programming variables appears in figure 3.12.

One other minor modification is necessary: *all* unexplored minimal extensions of maximal esperance of a true partial path P must be placed on the priority queue, not merely any unexplored minimal extension. This must be done to correctly handle the case of extensions which have equal *esperance.*

At first glance, it would appear that this procedure is sufficient to ensure that $f_i[g]$`.psi` is equal to $\psi^{g,\tau_{i-1}}$ when the best-first procedure attempts to extend $\{f_0, ..., f_{i-1}\}$ through f_i. In fact, this procedure underestimates $\psi^{g,\tau_{i-1}}$. Certainly if Q is a path terminating in g such that $d(Q) > \tau_{i-1}$ then the procedure forces $f_i[g]$.**psi** to contain ψ_Q, for then $\{Q, f_i\} \succ \{f_0, ..., f_i\}$ and hence Q was popped off the queue and extended through f_i before $\{f_0, ..., f_{i-1}\}$ was. However, there remains the case where $d(Q) = \tau_{i-1}$. In particular, consider the case where a number of paths of equal length conjoin at f_i, each path P_j terminating in g_j. Before any P_j can be extended through f_i, the field $f_i[g_j]$.**psi** must be updated for each g_j. But the field for an arbitrary g_j is not updated until the attempt is made to extend P_j, i.e., when we attempt to extend the first P_j none of the relevant fields have been updated.

The solution is to sum the viability functions for the P_j into $f_i[g_j].psi$ when we attempt to extend *any* of the P_j. The difficulty is in finding the P_j. We have the following.

Lemma 3.2.2 *Let Q, P be paths such that $d(Q) = d(P)$, and such that $\{P, f_i\}$ is an extension on the queue maximal under $\succ$. Then if $\{Q, f_i\}$ is unexplored, it is also on the queue and maximal under $\succ$.*

Proof: Note that $E(\{Q, f_i\}) = d(Q) + w(f_i) + D(f_i)$, and since $d(P) = d(Q)$, we have $E(\{Q, f_i\}) = E(\{P, f_i\})$. Further, since the paths $\{Q, f_i\}$ and $\{P, f_i\}$ have the same terminal node, the levels of their terminal nodes are equal. Hence $\{Q, f_i\} \not\succ \{P, f_i\}$ and $\{P, f_i\} \not\succ \{Q, f_i\}$. Hence if $\{Q, f_i\}$ is on the queue, it is maximal under $\succ$, for $\{P, f_i\}$ is maximal under $\succ$ by assumption. Since $\{Q, f_i\}$ is unexplored, then either it is on the queue or some prefix is. Every prefix Q' of $\{Q, f_i\}$ (excepting $\{Q, f_i\}$) is such that $Q' \succ \{Q, f_i\}$, and hence $Q' \succ \{P, f_i\}$. Hence Q' is not on the queue, for that would violate the assumption that $\{P, f_i\}$ is maximal under $\succ$. ∎

It is important to show that the converse holds as well.

Lemma 3.2.3 *Let Q, P be paths such that $\{P, f_i\}, \{Q, f_i\}$ are extensions on the queue maximal under $\succ$. Then $\{Q, f_i\}$, $\{P, f_i\}$ are both unexplored, and $d(Q) = d(P)$.*

Proof: Both P and Q are viable paths such that $\{P, f_i\}, \{Q, f_i\}$ are unexplored, by construction of the queue. Further, since both are maximal under $\succ$, we must have $E(\{Q, f_i\}) = E(\{P, f_i\})$, and hence

$$d(Q) + w(f_i) + D(f_i) = d(P) + w(f_i) + D(f_i)$$

hence $d(Q) = d(P)$ and done. ■

The picture the above two lemmas gives us is of a frontier of paths, maximal under $\succ$, each of whose viability functions must be summed into the relevant dynamic programming variable before any can be extended.

We now restate, operationally, how the field $f[g].psi$ is maintained for nodes f and g.

$$f[g].psi = \begin{cases} 0 & \text{initially} \\ f[g].psi + \psi_{\{P,f,g\}} & \{P, f, g\} \text{ is popped off the queue} \\ & \text{and found to be viable} \end{cases} \tag{3.1}$$

Let $\mathcal{Q}_{g,f_i}$ be the set of paths $\{Q, g\}$ such that $\{Q, g, f_i\}$ is an extension on the queue maximal under $\succ$. The computation of `viable_set`($\{P, f_i\}$) is then given by

$$\texttt{viable\_set}(\{P, f_i\}) = \sum_{U \subseteq S(f_i,P)} (S_U \frac{\partial f_i}{\partial f_{i-1}}) \prod_{g \in U} \left(f_i[g].psi + \sum_{Q \in \mathcal{Q}_{g,f_i}} \psi_Q \right) \tag{3.2}$$

We can now state the main result of this section, which proves the dynamic programming algorithm, detailed intuitively above and given in detail in figure 3.13, is correct.

Theorem 3.2.1 `viable_set`(f_i) $= \psi_P^{f_i}$

Proof: Induction on $\delta(f_i)$. For the base case, f_i is a primary input, whence $\frac{\partial f_i}{\partial f_{i-1}} = 1$ and done. Assume for $\delta(f_i) < L$. Now, for $\delta(f_i) = L$ the theorem holds if we can show that:

$$f_i[g].psi + \sum_{Q \in \mathcal{Q}_{g,f_i}} \psi_Q = \psi^{g,\tau_{i-1}}$$

for the general case. But:

$$\psi^{g,\tau_{i-1}} = \sum_{d(Q) \geq \tau_{i-1}} \psi_Q$$

where Q is a partial path ending in g. Now, if $d(Q) > \tau_{i-1}$, or if the level of g is less than that of the last node of P, then $Q \succ P$, and hence has been examined previously by the algorithm. Since $\delta(g) < L$, by the induction hypothesis for each such path Q ψ_Q was correctly calculated.

Further, as each such Q was popped off the path for extension through f_i, ψ_Q was added into $f_i[g].psi$ by equation (3.1) and hence $f_i[g].psi \supseteq \psi_Q$ for all such paths Q. There remains the case where Q and P are incomparable under $\succ$. In this case, by lemmas 3.2.2-3.2.3, Q and P are on the same frontier, and hence the viability function of Q is added into $f_i[g].psi$ before P is extended through f_i. Hence

$$f_i[g].psi + \sum_{d(Q) \geq \tau_i - 1} \psi_Q \supseteq \psi^{g, \tau_i - 1}$$

For equality, all we must show is that no path Q either

1. has had its viability function ψ_Q summed incorrectly into $f_i[g].psi$; or

2. is incorrectly in $\mathcal{Q}_{g,f_i}$

The first can only occur if the algorithm has examined some path Q before P with $d(Q) < d(P)$. But then $E(\{P, f_i\}) > E(\{Q, f_i\})$, contradiction. The second is forbidden by lemma 3.2.3, and done. ■

The code for the formal procedure is given in figure 3.13 and in figure 3.14. Much care is taken in this procedure to ensure that the scheduling assumptions of theorem 3.2.1 are met. In particular, when a path is found to be viable, *every* minimal extension of maximal esperance of that path is added to the queue; this is used to enforce the assumption that every unexplored path of maximal esperance terminating in minimal level (that is, every unexplored path maximal under the relation $\succ$) is present on the queue at all times; this is a requirement for lemmas 3.2.2-3.2.3 to hold. Similarly, when the last extension of a path of some fixed esperance is found to be viable, or not, then the set of extensions of that path of the next higher esperance must be added to the queue.

While this algorithm finds every viable path of maximal length, it does so at great expense. Recall that the relation $\succ$ required that we break ties in esperance in favour of the path whose terminus is of lesser level. This forced the best-first procedure ordered on esperance into what is almost a *breadth-* first procedure, and leaves us open to the possibility that a larger number of paths will be explored than really need to be. In fact, if there are K_1 paths longer than the longest true path, and K_2 paths as long as the longest true path, this procedure explores $O((K_1 + K_2)D)$ paths, as opposed to the $O(K_1 D)$ paths that strictly need to be explored.

```
find_longest_true_path(){
    Initialize queue to primary inputs of the circuit
    while(queue ≠ nil) {
        frontier ← set of paths on queue maximal under ≻;
        foreach extension (path,g) on frontier {
            k is the last node of path;
            g[k].psi ← g[k].psi + ψ(path);
        }
        while(((path, g) ← pop(frontier)) ≠ nil) {
            ψ ← viability_function(path, g)
            if(ψ ≢ 0) {
                new_path ← {path, g};
                if(g is an output) return new_path;
                ψ(new_path) ← ψ;
                foreach extension np ← {new_path, h} of new_path
                    if(E(np) = E(new_path))
                        insert np on queue
            }
            if every extension ext of path s.t.
                E(ext) = E(new_path) has been explored {
                ext1 is next best extension of path
                push ext1 on queue;
                foreach extension ext2 of path with E(ext1) = E(ext2)
                    push ext2 on queue;
            }
        }
    }
}
```

Figure 3.13: Dynamic Programming Procedure to Find the Longest Viable Path

```
viability_function(path, g) {
    k is the last node of path;
    sense_fn ← 0;
    esp = E({path, g});
    foreach subset U of the side fanins of g {
        product ← S_U ∂g/∂k;
        foreach j ∈ U while product ≢ 0 {
            sum ← g[j].psi;
            product ← product * sum;
        }
        sense_fn ← sense_fn + product;
    }
    return ψ(path) * sense_fn;
}
```

Figure 3.14: Viability Function for Dynamic Programming Procedure

What we would like is some way to break ties in favour of the path whose terminus is of *greater* level. We can do this if we guarantee that if we reject any viable path incorrectly, we will subsequently accept one at least as long. This is obtained as a consequence of the following theorem and corollary: in a symmetric network, if there are a set of partial paths of equal length conjoined at a single node and if one may be extended to a full path, then all may. This theorem also shows that we need not add the sum of the viability functions of the maximal paths on the queue into the field $f[g].psi$ when computing the viable set for extending a path from g through f. We now prove this statement.

Theorem 3.2.2 *Consider a symmetric network. Let P_1, P_2 be partial paths, viable under c, with $d(P_1) = d(P_2)$, P_1 terminates in g_1, P_2 terminates in g_2, g_1 and g_2 are fanins to a node h, such that $\{P_1, h\}$ is viable under c. Then $\{P_2, h\}$ is viable under c.*

Proof: Suppose $g_1 \neq g_2$. Since $\{P_1, h\}$ viable under c, there is some set

U of the fanins of h such that:

$$c \subseteq \mathcal{S}_U \frac{\partial h}{\partial g_1} \prod_{g \in U} \psi^{g,d(P_1)}$$

since there is a path terminating in g_2 of length $d(P_1)$ (namely P_2), $c \subseteq \psi^{g_2,d(P_1)}$. Hence we can choose $g_2 \in U$, for certainly if $g_2 \notin U$:

$$c \subseteq \mathcal{S}_{g_2} \mathcal{S}_U \frac{\partial h}{\partial g_1} \prod_{g \in U} \psi^{g,d(P_1)} \psi^{g_2,d(P_1)} = \mathcal{S}_{U+\{g_2\}} \frac{\partial h}{\partial g_1} \prod_{g \in U+\{g_2\}} \psi^{g,d(P_1)}$$

Since, by symmetry, $\mathcal{S}_{U-\{g_2\}+\{g_1\}} \frac{\partial h}{\partial g_2} = \mathcal{S}_U \frac{\partial h}{\partial g_1} \supseteq c$, and since $\psi^{g_1,d(P_2)} \supseteq c$, we have:

$$c \subseteq \mathcal{S}_{U-\{g_2\}+\{g_1\}} \frac{\partial h}{\partial g_2} \prod_{g \in U-\{g_2\}+\{g_1\}} \psi^{g,d(P_2)}$$

and so $\{P_2, h\}$ is viable under c. Now, if $g_1 = g_2$, and we have the set U such that that:

$$c \subseteq \mathcal{S}_U \frac{\partial h}{\partial g_1} \prod_{g \in U} \psi^{g,d(P_1)}$$

since $g_1 = g_2$ and $d(P_1) = d(P_2)$, a simple substitution shows that:

$$c \subseteq \mathcal{S}_U \frac{\partial h}{\partial g_2} \prod_{g \in U} \psi^{g,d(P_2)}$$

and hence $\{P_2, h\}$ is viable under c. ■

Corollary 3.2.3 *In a symmetric network, if $\{P_1, .., P_n\}$ are partial paths, viable under c, each terminating in a fanin to some node h_0, such that $d(P_i) = d(P_j)$ for all i, j, and if one of the P_i is a prefix (through h) to a path $Q_i = \{P_i, h_0, h_1, ..., h_r\}$, viable under c, then* each *P_j is a prefix through h to a path $Q_j = \{P_j, h_0, h_1, ..., h_r\}$, viable under c, such that $d(Q_i) = d(Q_j)$.*

Proof: Induction on r. If $r = 0$, immediate from theorem 3.2.2. Assume for $r < R$. If $r = R$, let the P_1 of theorem 3.2.2 be $\{P_i, h_0, ...h_{r-1}\}$, and h of that theorem be h_R. Result follows immediately. ■

This gives us the tool we need. We can only incorrectly reject a path if it is of the form P_i in the above the theorem. However, of the set $\{P_1, .., P_n\}$, we will examine *one* last, and if we have accepted no other prior to that we shall accept that one. Note that this will not give us the list of *all* viable paths of delay equal to the delay of the longest viable path; that must be accomplished by a close variant of the original algorithm, detailed below.

The improved dynamic programming procedure is given in figure 3.15. Note that this procedure is much closer in form to the best-first procedure demonstrated above.

```
find_longest_true_path(){
    Initialize queue to primary inputs of the circuit
    while(((path, g) ← pop(queue)) ≠ nil) {
        k is the last node of path;
        g[k].psi ← g[k].psi + ψ(path);
        ψ ← viability_function(path, g)
        if(ψ ≢ 0) {
            new_path ← {path, g};
            if(g is an output) return new_path;
            ψ(new_path) ← ψ;
            np ← {new_path, h} is best extension of new_path
            insert np on queue
        }
        ext ← {path, f} is the next best extension of path
    if ext is not nil push ext on queue;
    }
}
```

Figure 3.15: Improved Dynamic Programming Procedure to Find an LVP

The correctness of this procedure is easy to establish.

Theorem 3.2.4 *Let $P = \{f_0, ..., f_m, h_0, ..., h_p\}$ be a viable path under c. Then either P is reported as viable by the algorithm, or some path $Q = \{g_0, ..., g_n, h_0, ..., h_p\}$, with $g_n \neq f_m$ and $d(\{f_0, ..., f_m\}) = d(\{g_0, ..., g_n\})$ is reported as viable by the algorithm.*

```
viability_function(path, g) {
    k is the last node of path;
    sense_fn ← ∂g/∂k;
    esp = E({path, g});
    foreach j ∈ S(g, k) {
        sense_fn ← (1 + S_j) * sense_fn;
    }
    return ψ(path) * sense_fn;
}
```

Figure 3.16: Viability Function for Improved Dynamic Programming Procedure

Proof: If $P = \{f_0, ..., f_m, h_0, ..., h_p\}$ is rejected by the algorithm, then at some node the computed viability function of the path as computed is a proper subset of the viability function of the partial path to that node. Let h_0 be the first such node. The viability function to that point is

$$\psi_P^{f_m}[\sum_U \mathcal{S}_U \frac{\partial h_0}{\partial f_m} \prod_{g \in U} \psi^{g, d(\{f_0, ..., f_m\})}]$$

Now, by assumption, $\psi_P^{f_m}$ has been correctly calculated by the procedure, and hence we must have that

$$[\sum_U \mathcal{S}_U \frac{\partial h_0}{\partial f_m} \prod_{g \in U} \psi^{g, d(\{f_0, ..., f_m\})}]$$

is undercomputed. Hence

$$\psi^{g_i, d(\{f_0, ..., f_m\})} \supseteq h_0\texttt{[}g_i\texttt{].psi}$$

for some set of side inputs g_i, and each $g_i \neq f_m$. Hence there exist viable paths, terminating in one of the g_i, of length $\geq d(\{f_0, ..., f_m\})$ that had not been traced when P was extended. Each path of length $> d(\{f_0, ..., f_m\})$ had been traced, and hence each such path is of length $d(\{f_0, ..., f_m\})$. Let $Q_j = \{g_{0j}, ..., g_i\}$ be the set of such paths. Now, by theorem 3.2.2 one of the paths $\{Q_j, h_0, ..., h_p\}$ is the long viable path reported by the algorithm. This path satisfies the assumptions of the path Q of the theorem, and so done. ∎

A final improvement can be made to the calculation of the viability function. Note that the procedure as given in figure 3.14 requires the computation of the viability function for each subset of the n side inputs to **g**. Hence this algorithm is exponential in the number of fanins to any node. Now, if the gates are of bounded fanin, this is not a major concern, but it behooves us to consider the most efficient manner of computing this sum.

Now, note that if the dynamic programming variables are all nonzero, there is no help for it: the result is exponential in the number of fanins of the gate. In practice, however, this is a pathological case. We are interested in speeding up the *general* case, where most of the dynamic programming variables are identically 0.

The equation implemented by the viability function procedure is (2.3). Using operator multiplication ($\mathcal{S}_{xy}f = \mathcal{S}_x\mathcal{S}_y f$), we can rewrite equation (2.3) as follows:

$$\psi_P^{f_i} = \left[\prod_{g \in S(f_i,P)} (1 + \mathcal{S}_g \psi^{g,\tau_{i-1}})\right] \frac{\partial f_i}{\partial f_{i-1}} \tag{3.3}$$

Note that this equation requires $O(2n)$ operations, as opposed to $O(2^n)$ for equation (2.3), and it is a trivial induction to demonstrate that the two equations compute the same function. It is also possible to demonstrate that the worst-case size of (3.3) must be $O(2^r)$, where r is the number of nonzero dynamic programming variables. It is easy to modify 3.14 to compute (3.3), and the resulting code is in figure 3.16.

3.3 Dynamic Programming Algorithm Example

We now demonstrate an example of the dynamic programming procedure in action.

Consider the circuit of figure 2.3. We analyze this circuit using the algorithm of figure 3.15 and figure 3.16. For this circuit, we have $\frac{\partial z}{\partial x} = \frac{\partial z}{\partial y} = 0$, $\frac{\partial x}{\partial u} = \frac{\partial y}{\partial a} = \overline{a}$, $\frac{\partial x}{\partial w} = \frac{\partial y}{\partial v} = a$, $\frac{\partial u}{\partial a} = \frac{\partial y}{\partial a} = 1$. Since each gate in this

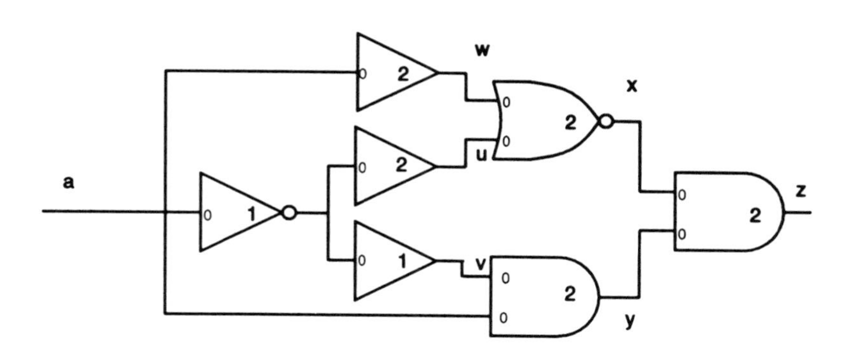

Figure 3.17: Circuit on Algorithm Entry

circuit is one or two input, note that the viability equations reduce to

$$\psi_P^{f_i} = \frac{\partial f_i}{\partial f_{i-1}} + \psi^{g,\tau_{i-1}}$$

We represent the variable `g[k].psi` at the `k` input connection of gate `g`. For example, in figure 3.17, the variable `x[w].psi` is the 0 beside the upper connection in the NOR gate. The value of each such variable is 0 at the entry of the algorithm, as indicated in the figure. For convenience we have explicitly represented the delay buffers in this example.

The algorithm first attempts to extend the path a through u, ORing the viability function of the path $\{a\}$ (1) into the variable `u[a].psi`[2]; as the extension through u succeeds with viability function 1, the algorithm sets `x[u].psi` = 1. The extension through x succeeds with viability function $\overline{a}$, and `z[x].psi` is set to $\overline{a}$. The extension through z fails, and the circuit now appears as in figure 3.18.

The algorithm now explores the path $\{a, v, y, z\}$. Again, $\{a, v, y\}$ is a viable path with viability function a. The algorithm now attempts to extend y to z. The function $\psi^z_{\{a,v,y\}}$ is $\frac{\partial z}{\partial y} + \bar{a}$; hence $\psi_{\{a,v,y,z\}} = a(0+\bar{a}) =$ 0, and the extension fails, leaving the circuit in figure 3.19.

The algorithm now explores the path $\{a, w, x, z\}$. $\{a, w, x\}$ is a viable path with viability function 1. The algorithm now attempts to extend x

[2]for convenience, `u[a].psi` and `v[a].psi` are represented in the same variable

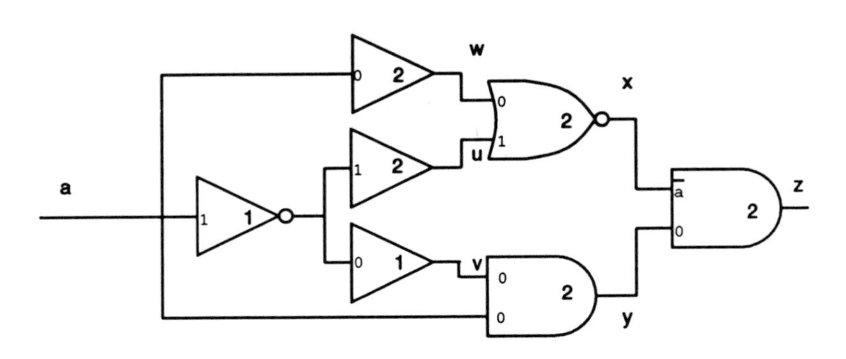

Figure 3.18: Circuit After {a,u,x,z} Explored

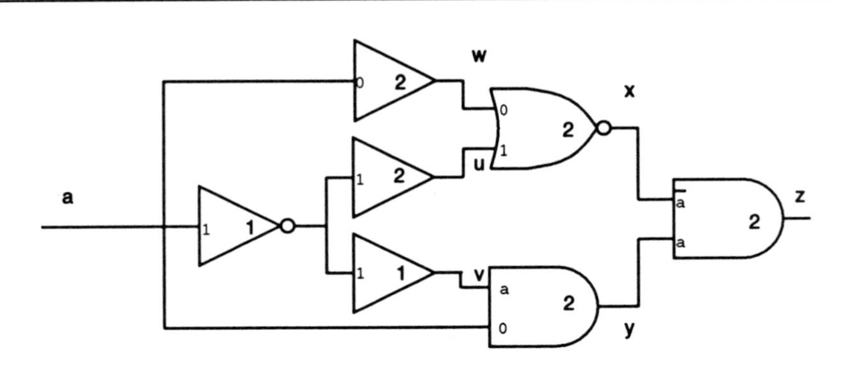

Figure 3.19: Circuit After {a,v,y,z} Explored

to z. The function $\psi^z_{\{a,w,x\}}$ is $\frac{\partial z}{\partial x} + a$; hence $\psi_{\{a,w,x,z\}} = 1(0+a) = a$, and the extension succeeds, reporting the longest viable path as $\{a, w, x, z\}$, a path of length 6. The final circuit is as appears in figure 3.20

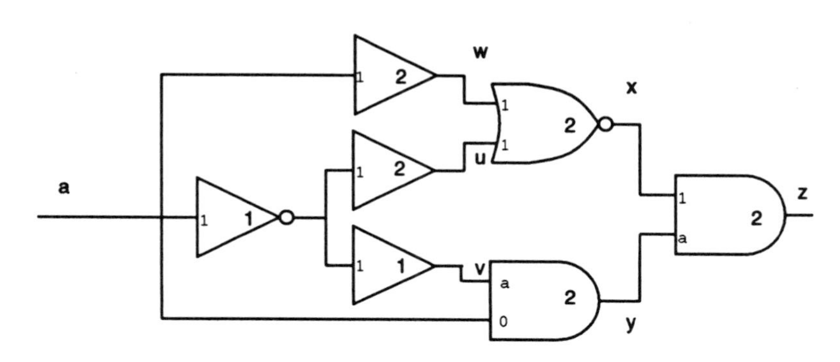

Figure 3.20: Circuit After {a,v,y,z} Explored

3.4 Finding all the Longest Viable Paths

The improved dynamic programming procedure described in the preceding sections has the flaw that only one of the longest viable paths is reported; if all the longest viable paths are to be reported, then the original dynamic programming procedure is preferred. As mentioned above, we often wish a procedure which returns all the longest paths at or above a given length. This procedure can be easily obtained by a simple modification of the first dynamic programming procedure, given in figure 3.21.

This procedure returns all viable paths of length $\geq T$, for some T. T is either directly set, or is chosen to be some ϵ less than the length of the longest viable path; hence choosing $\epsilon = 0$ finds all the longest viable paths.

If search-by-ϵ is selected, the initial threshold is set to 0; this simply ensures that no path will fail to meet the threshold test. When a viable path is found, if the threshold is still 0 then this is the first viable path

```
find_longest_true_path(){
    Initialize queue to primary inputs of the circuit
    long_paths = ∅;
    if search_by_epsilon T = 0;
    while((((path, g) ← pop(queue) ) ≢ nil) and (E(path, g) ≥ T)) {
        k is the last node of path;
        g[k].psi ← g[k].psi + ψ(path);
        ψ ← viability_function(path, g)
        if(ψ ≢ 0) {
            new_path ← {path, g};
            if(g is an output) {
                if (search_by_epsilon and T = 0)
                    T = length(new_path) - ε;
                long_paths = long_paths ⋃ {g};
            }
            else {
                ψ(new_path) ← ψ;
                foreach extension np ← {new_path, h} of new_path
                    if(E(np) = E(new_path))
                        insert np on queue
            }
        }
        if every extension ext of path s.t.
            E(ext) = E(new_path) has been explored {
            ext1 is next best extension of path
            push ext1 on queue;
            foreach extension ext2 of path with E(ext1) = E(ext2)
                push ext2 on queue;
        }
    }
return long_paths;
}
```

Figure 3.21: Dynamic Programming Procedure to Find All the Long Viable Paths

found; the threshold is set to the length of this path $-\epsilon$.

Whenever a full viable path is found, it is added to `long_paths`. By construction, this path is viable and is of length $\geq T$. When no more paths of potential full length $\geq T$ remain to be explored, the algorithm terminates and `long_paths` is returned.

By construction it is easy to see that only viable paths of length $\geq T$ are returned. To see that all such paths are returned, the similarity of this procedure and the procedure of figure 3.13, together with theorem 3.2.1 suffice.

The complexity of this algorithm is $O(KDS \log D)$, where D is the diameter of the graph, S is the cost of a SAT call, and K is the number of paths, false and true, of length $\geq T$. It is easy to see that this is the minimum number of paths which must be examined by any procedure which purports to solve this problem.

3.5 Recent Work

In work yet to appear in the conference and archival literature, the calculation of the sensitization functions and their justification has been subject to significant attention. Initial implementations of the algorithms given here led to sensitization functions at each node which were very large. Examination of the viability equations demonstrated that the power-set sum given in the definition of $\psi_P^{f_i}$ was the culprit. In [61], careful attention was given to the calculation of these functions. It was demonstrated that by factorization of the power-set sum for $\psi_P^{f_i}$ and by careful representation of $\psi^{g,t}$, each sensitization function at a connection could be represented in size linear in the number of fanins to the output gate of the connection plus the number of false paths longer than the longest true path through the connection.

Rewriting each function so that the representation was small was only half the battle, however. In addition, the functions must be tested for 0 efficiently. Two methods were used. First, Boolean Decision Diagrams[19] were formed for each function; such diagrams may be very large, but have the great advantage that the only representation for a function equivalent to 0 is the diagram representing 0. Second, a new procedure for Boolean Satisfiability due to T. Larrabee was used [55]. This procedure, which exploits the satisfiability don't-care set and the fact that satisfiability is linear time given clauses with only two variables, has been found effective in test-generation applications. In [61], it was reasoned that this technique should be useful for path justification.

In all but one of the ISCAS benchmarks, Larrabee's procedure with the recursive function formulation for the sensitization functions yielded a solution in a few tens of seconds on a DECstation 3100. The exception to this general rule was the benchmark C6288, a multiplier. This particular example has over 40,000 false paths of length equal to the length of the longest path. Further, the longest true path is also of length equal to this length. The results given in the following tables, compare the CPU times and delay estimates given by static sensitization (known to provide an underestimate of delay), by the viability criterion, and by the Brand-Iyengar criterion[11], which, as we shall see in chapter 4, provides an estimate no less than that of the viability criterion.

Name	Delay Estimate				CPU Time (secs)		
	Longest	Static	B-I	Viable	Static	B-I	Viable
C880	24.0	24.0	24.0	24.0	6	8	8
C1355	24.0	24.0	24.0	24.0	11	14	13
C1908	40.0	39.0	39.0	39.0	30	35	44
C2670	32.0	30.0	31.0	30.0	143	70	5868
C3540	47.0	46.0	46.0	46.0	64	113	123
C5315	49.0	47.0	47.0	47.0	53	68	141
C6288	124.0	*	*	*	*	*	*
C7552	43.0	42.0	42.0	42.0	84	96	323

Table 3.1: Timing analysis using BDD's on unoptimized ISCAS circuits

Name	Delay Estimate				CPU Time (secs)		
	Longest	Static	B-I	Viable	Static	B-I	Viable
C880	24.0	24.0	24.0	24.0	2	2	3
C1355	24.0	24.0	24.0	24.0	4	4	3
C1908	40.0	39.0	39.0	39.0	97	96	175
C2670	32.0	31.0	32.0	32.0	135	20	340
C3540	47.0	46.0	46.0	46.0	60	56	89
C5315	49.0	47.0	47.0	47.0	24	17	82
C6288	124.0	124.0	124.0	124.0	≈20 hours		
C7552	43.0	42.0	42.0	42.0	18	18	25

Table 3.2: Timing analysis using SAT on unoptimized ISCAS circuits

Name	Delay Estimate				CPU Time (secs)		
	Longest	Static	B-I	Viable	Static	B-I	Viable
5xp1	11.0	9.0	9.0	9.0	2	2	4
bw	29.0	25.0	26.0	25.0	13	17	5328
des	15.0	13.0	13.0	13.0	97	96	175
misex1	9.0	7.0	7.0	7.0	35	35	229
rot	19.0	18.0	18.0	18.0	13	12	33

Table 3.3: Timing analysis using SAT on select optimized MCNC circuits

Chapter 4

System Considerations and Approximations

The theory and algorithms described to this point capture the nature of the problem. Nevertheless, the viability procedure is theoretically expensive; the inner loop of the procedure is a general satisfiability problem, and hence is strongly suspected to be of exponential complexity. Further, there is some bookkeeping associated with the computation of the viability function. In general, some applications may prefer a faster answer and a poorer approximation to the longest viable path, so long as the assurance is given that such an approximation will not *underestimate* the length of the longest viable path. In this section we explore such performance/quality tradeoffs, and end by giving a polynomial approximation to viability.

4.1 Approximation Theory and Practice

The algorithms developed in the previous chapter removed the unnecessary inefficiencies in computing the viable set. However, the necessity of computing a sum over the power set of the inputs at each node remains a potentially expensive operation. This operation appears unavoidable for exact computation of the viable sets for the network. However, approximate solutions may be found. Of course, these approximations must be conservative: each approximation, to be valid, must be shown to upper bound the delay down the longest (dynamically) sensitizable path, and must be shown to obey monotone speedup. In this section we explore such approximation techniques.

In previous chapters, we have noted the duality of network or path conditions and some boolean functions, to which we should give a name; let us call these *path logic functions.* Since every algorithm which attempts to solve the false path problem explicitly or implicitly associates a logic function Γ_P with each path P and computes its satisfying set, it seems that one way to validate (or not) such algorithms is to consider the properties of its associated family of logic functions. The most clearly interesting property is whether or not the function is *correct* in the obvious sense: if one accepts the critical delay as the longest path P such that Γ_P is satisfiable, then no longer path P' may transmit an event in this or any faster network. We formalize this in the following definition.

Definition 4.1.1 *A family of path logic functions Γ_P on a boolean network N is said to be* **critical path correct** *iff the longest path P such that Γ_P is satisfiable is longer than the longest dynamically sensitizable path in any network N', where N' is obtained from N by reducing some or none internal delays.*

Our familiar properties of dynamic sensitizability and monotone speedup may be expressed as properties of path logic functions:

Definition 4.1.2 *A family of path logic functions Γ_P on a boolean network N is said to have the* **dynamic sensitization property** *iff Γ_P is satisfiable for every dynamically sensitizable path P.*

Definition 4.1.3 *A family of path logic functions Γ_P on a boolean network N is said to have the* **monotone speedup property** *iff for each satisfiable $\Gamma_{P'}$ in a sped-up network N' there is a path P in N such that $d_N(P) \geq d_{N'}(P')$ and Γ_P is satisfiable.*

These definitions may seem like old wine in new bottles, inasmuch as these are merely near-repetitions of our earlier definitions. However, most find old wine highly palatable, and occasionally new bottles permit us to see the body more clearly. We will try to peer through the new clear glass now.

One fact that is immediately apparent is that the two enumerated properties on a family of path logic functions are *not* required for critical path correctness. While it is true that monotone speedup and dynamic sensitization together imply critical path correctness (as we have seen), the converse is not true. The following theorem gives us the family of counterexamples.

Theorem 4.1.1 (Approximation Theorem) *Let Γ_P be a critical path correct family of path logic functions. Every family of path logic functions $\widehat{\Gamma_P}$ .s.t. at least one of the following holds:*

1. *$\widehat{\Gamma_P} \supseteq \Gamma_P$, for every P or*

2. *for every satisfiable P s.t. Γ_P is satisfiable there is a P' with $d(P) \leq d(P')$ s.t. $\widehat{\Gamma_{P'}}$ is satisfiable*

is also critical path correct.

Proof: The proof is almost a triviality. If $\widehat{\Gamma_P} \supseteq \Gamma_P$ for every P, and Q' is the longest sensitizable path in any of the family of networks (Q' is sensitizable in some N'), then we must have some Q s.t. $d_N(Q) \geq d_{N'}(Q')$ and Γ_Q is satisfiable. Since $\widehat{\Gamma_Q} \supseteq \Gamma_Q$, we have that $\widehat{\Gamma_Q}$ is satisfiable, and so $\widehat{\Gamma_P}$ is critical path correct. For the second item, we have Q, N, Q', N' and Γ_Q satisfiable as before. Since Γ_Q is satisfiable, there is Q'' s.t. $d(Q'') \geq d(Q)$ and $\widehat{\Gamma_{Q''}}$ is satisfiable, and since $d_N(Q'') \geq d_N(Q) \geq d_{N'}(Q')$ we have that $\widehat{\Gamma_P}$ is critical path correct. ■

Examples of functions which are critical path correct but which do not have one, or both, of the two critical properties abound. For example, the (trivial) family of functions

$$\Xi_P = \begin{cases} 1 & P \text{ is the longest path in the network} \\ 0 & \text{otherwise} \end{cases}$$

obviously does not have the dynamic sensitization property, but is critical path correct. Similarly, it seems likely that there are families of functions which do not obey monotone speedup but which are critical path correct.

On the other hand, function families which do not satisfy the two properties but which are critical path correct seem to be fated to be relatively weak upper bounds, unless there is some further theory to be discovered here. The only tool one has for proving such a function family correct is theorem 4.1.1, for which one must have shown the existence of another correct function family which yields a better result, and so one can always do better by computing the latter. This theorem has its uses, however. First, we may analyze and prove correct existing algorithms. Second, though we have no rigorous proof, this theorem and its associated discussion leads us to the belief that one can do no better than viability. Third, the viability function is expensive to compute, since it involves computing a sum over the power set of the side inputs at every node; indeed, even if the sum-of-products expression for the viability function at each node involved only two terms, it is easy to see that the function on some path P could grow as large as 2^D, where D is the diameter of the graph. This theorem permits us to consider other, easier-to-compute functions which satisfy the assumptions of theorem 4.1.1. We turn to the first of these, an analysis of the weak viability procedure.

4.2 "Weak" Viability

Recall the viability equations (2.2)-(2.4)

$$\begin{aligned}
\psi_P &= \prod_{i=0}^{m} \psi_P^{f_i} \\
\psi_P^{f_i} &= \sum_{U \subseteq S(f_i,P)} (\mathcal{S}_U \frac{\partial f_i}{\partial f_{i-1}}) \prod_{g \in U} \psi^{g,\tau_i-1} \\
\psi^{g,t} &= \sum_{Q \in \mathcal{P}_g, d(Q) \geq t} \psi_Q
\end{aligned}$$

The heart of the definition is equation (2.3)

$$\psi_P^{f_i} = \sum_{U \subseteq S(f_i,P)} (\mathcal{S}_U \frac{\partial f_i}{\partial f_{i-1}}) \prod_{g \in U} \psi^{g,\tau_i-1}$$

The power-set sum in this equation is also the heart of the inefficiencies inherent in the definition. The expression cannot be immediately simplified, for no term of

$$\sum_{U \subseteq S(f_i,P)} (\mathcal{S}_U \frac{\partial f_i}{\partial f_{i-1}}) \prod_{g \in U} \psi^{g,\tau_i-1}$$

inherently covers any other. Indeed, for $V \subset U$, we have:

$$\mathcal{S}_V \frac{\partial f_i}{\partial f_{i-1}} \subseteq \mathcal{S}_U \frac{\partial f_i}{\partial f_{i-1}}$$

but also:

$$\prod_{g \in V} \psi^{g,\tau_{i-1}} \supseteq \prod_{g \in U} \psi^{g,\tau_{i-1}}$$

However, if we weaken the definition of viability slightly:

Definition 4.2.1 *The* **weak viability function** *of a path*

$$P = \{f_0, ..., f_m\}$$

is defined as:

$$\varphi_P = \prod_{i=0}^{m} \varphi_P^{f_i} \tag{4.1}$$

where

$$\varphi_P^{f_i} = \sum_{U \subseteq S(f_i,P)} (\mathcal{S}_U \frac{\partial f_i}{\partial f_{i-1}}) \prod_{g \in U} \varphi^{g,\tau_{i-1}} \tag{4.2}$$

and

$$\varphi^{g,t} = \begin{cases} 0 & \sum_{Q \in \mathcal{P}_{g,t}} \varphi_Q = 0 \\ 1 & otherwise \end{cases} \tag{4.3}$$

Paths P such that $\varphi_P \not\equiv 0$ are said to be weakly viable.

We have immediately:

Theorem 4.2.1 *For every path P, $\varphi_P \supseteq \psi_P$.*

Proof: Let $P = \{f_0, ..., f_m\}$. Induction on $\delta(f_m)$. If f_m is a primary input, trivial. Suppose $\varphi_Q \supseteq \psi_Q$ for paths $Q = \{g_0, ..., g_n\}$ with $\delta(g_n) < N$. If $\delta(f_m) = N$, the result holds if we can show:

$$\varphi_P^{f_m} \supseteq \psi_P^{f_m}$$

But this is immediate, for we have that:

$$\varphi_P^{f_m} = \sum_{U \subseteq S(f_m,P)} (\mathcal{S}_U \frac{\partial f_m}{\partial f_{m-1}}) \prod_{g \in U} \varphi^{g,\tau_{m-1}}$$

Since by induction we have that:

$$\varphi_Q \supseteq \psi_Q$$

for each Q terminating in a fanin of f_m, we must have that:

$$\sum_{Q \in \mathcal{P}_g, t} \varphi_Q \supseteq \sum_{Q \in \mathcal{P}_g, t} \psi_Q$$

Hence if $\psi^{g,\tau_{m-1}} \not\equiv 0$ we must have $\varphi^{g,\tau_{m-1}} = 1$, whence $\varphi^{g,\tau_{m-1}} \supseteq \psi^{g,\tau_{m-1}}$ for each g, and hence we may write:

$$\varphi_P^{f_m} \supseteq \sum_{U \subseteq S(f_m, P)} (\mathcal{S}_U \frac{\partial f_m}{\partial f_{m-1}}) \prod_{g \in U} \psi^{g,\tau_{m-1}}$$

and the right-hand side is obviously $\psi_P^{f_m}$, whence the result. ■

This serves to show that φ is a critical-path correct path function on symmetric networks, by the approximation theorem. The attractive thing about φ is the following observation:

Theorem 4.2.2 *Let $P = \{f_0, ..., f_m\}$. Let V be the maximal set of g s.t. $\varphi^{g,\tau_{i-1}} = 1$. Then*

$$\mathcal{S}_V \frac{\partial f_i}{\partial f_{i-1}} = \varphi_P^{f_i}$$

Proof: From (4.2):

$$\varphi_P^{f_i} = \sum_{U \subseteq S(f_i, P)} (\mathcal{S}_U \frac{\partial f_i}{\partial f_{i-1}}) \prod_{g \in U} \varphi^{g,\tau_{i-1}}$$

Now, we have that

$$\mathcal{S}_V \frac{\partial f_i}{\partial f_{i-1}} \prod_{g \in V} \varphi^{g,\tau_{i-1}}$$

is a term of this series, and since $\varphi^{g,\tau_{i-1}} = 1$ for all $g \in V$ this is

$$\mathcal{S}_V \frac{\partial f_i}{\partial f_{i-1}}$$

whence

$$\mathcal{S}_V \frac{\partial f_i}{\partial f_{i-1}} \subseteq \varphi_P^{f_i}$$

Further, consider an arbitrary term of (4.2), say:

$$(\mathcal{S}_U \frac{\partial f_i}{\partial f_{i-1}}) \prod_{g \in U} \varphi^{g,\tau_{i-1}}$$

Now, if $U \subseteq V$, this is certainly $\subseteq \mathcal{S}_V \frac{\partial f_i}{\partial f_{i-1}}$. However, if $U \supset V$, then by construction there is some $g \in U$ such that $\varphi^{g,\tau_{i-1}} = 0$, by choice of V. Hence we must have that:

$$(\mathcal{S}_U \frac{\partial f_i}{\partial f_{i-1}}) \prod_{g \in U} \varphi^{g,\tau_{i-1}} = 0$$

and so

$$(\mathcal{S}_U \frac{\partial f_i}{\partial f_{i-1}}) \prod_{g \in U} \varphi^{g,\tau_{i-1}} \subseteq \mathcal{S}_V \frac{\partial f_i}{\partial f_{i-1}}$$

and hence:

$$\mathcal{S}_V \frac{\partial f_i}{\partial f_{i-1}} \supseteq \varphi_P^{f_i}$$

and therefore:

$$\mathcal{S}_V \frac{\partial f_i}{\partial f_{i-1}} = \varphi_P^{f_i}$$

■

In other words, only a single term of the series (4.2) (the last nonzero term) need be taken. While this can still lead to an exponential blowup in the size of the function, it seems far less likely. Further, under some circumstances one can guarantee that the size of φ_P will remain bounded (in fact, one can guarantee that φ_P will consist of a single cube). We detail these circumstances here.

Theorem 4.2.3 *Let N be any network such that for any node f, any input x of f, $\frac{\partial f}{\partial x}$ is a single cube. Then for each path P through the network, φ_P is a single cube*

Proof: For the proof, we note that:

$$\varphi_P = \prod_{i=0}^{m} \varphi_P^{f_i}$$

where

$$\varphi_P^{f_i} = \sum_{U \subseteq S(f_i,P)} (\mathcal{S}_U \frac{\partial f_i}{\partial f_{i-1}}) \prod_{g \in U} \varphi^{g,\tau_{i-1}}$$

now, from theorem 4.2.2, we know that this last can be written

$$\varphi_P^{f_i} = \mathcal{S}_V \frac{\partial f_i}{\partial f_{i-1}}$$

Where V be the maximal set of g s.t. $\varphi^{g,\tau_{i-1}} = 1$. Further, for any function h, any set of inputs V, if h is a single-cube function then so is $S_V h$, hence $\varphi_P^{f_i}$ is a single-cube function for each f_i. The product of single-cube functions is a single-cube function, and so done. ■

Weak viability is an attractive alternative to viability when computation time is at a premium, and can be used in conjunction with other approximation techniques. For example, the weak-viability function is therefore a single cube for a network consisting only of NOR, NAND, OR,

AND and NOT gates, or some subset thereof. One can always transform *any* network into such a network, through a variety of macroexpansion transformations, and guarantee correctness through theorem 2.5.1; one can then use weak viability on the transformed network.

It is interesting to formally define the set of paths P for which $\varphi_P \neq 0$. We give these here.

Definition 4.2.2 *A path $P = \{f_0, ..., f_m\}$ is said to be* **weakly viable** *under an input cube c if, at each node f_i there exists a (possibly empty) set of side inputs $U = \{g_1, ..., g_n\}$ to P at f_i, such that, for each j,*

1. *g_j is the terminus of a path Q_j,*
2. *$d(Q_j) \geq \tau_{i-1}$ and Q_j is weakly viable*
3. *$(\mathcal{S}_U \frac{\partial f_i}{\partial f_{i-1}})(c) = 1$*

Notice that the definition of weakly viable paths differs only marginally from that of viable paths; the principle difference is that no record is kept of the cube under which the side path is viable (item 2).

We must formally show the equivalence of the function and the definition.

Theorem 4.2.4 *$P = \{f_0, ..., f_m\}$ is weakly viable under some minterm c iff c satisfies φ_P*

Proof: $\Longrightarrow$ $P = \{f_0, ..., f_m\}$ is weakly viable under c. Induction on $\delta(f_m)$. The base case is trivial, so assume for $\delta(f_m) < L$. Let $\delta(f_m) = L$. We must show that for each f_i, $c \in \varphi_P^{f_i}$. Now, if $c \in \frac{\partial f_i}{\partial f_{i-1}}$, done. Otherwise, since the path is weakly viable under c, then we must have that there is a subset $U = \{g_1, ..., g_k\}$ of $S(f_i, P)$, where each g_j terminates a path Q_j with $d(Q_j) \geq \tau_{i-1}$ and Q_j is weakly viable. By induction, $\varphi_{Q_j} \neq 0$. Hence $\varphi^{g_j, \tau_{i-1}} = 1$ for every j by 4.3. Further, $c \in \mathcal{S}_U \frac{\partial f_i}{\partial f_{i-1}}$, and done.
$\Longleftarrow$ $c \in \varphi_P$. Induction on $\delta(f_m)$. The base case is trivial, so assume for $\delta(f_m) < L$. Let $\delta(f_m) = L$. We must show that the definition of weak viability holds for each f_i. Now, if $c \in \frac{\partial f_i}{\partial f_{i-1}}$, done. Otherwise, we must show that there exists a set of side inputs U meeting the conditions of the definition of weak viability with $\mathcal{S}_U \frac{\partial f_i}{\partial f_{i-1}} \supseteq c$. Since $c \in \varphi_P^{f_i}$, then we must have that there is a subset $U = \{g_1, ..., g_k\}$ of the side inputs,

and for each g_j, $\varphi^{g_j,\tau_{i-1}} = 1$. Now, by the definition of $\varphi^{g_j,\tau_{i-1}}$, we must have for each j there must be a path Q_j terminating in g_j weakly viable with $d(Q_j) \geq \tau_{i-1}$. $\delta(g_j) < L$, so by induction Q_j is weakly viable, and $c \in \mathcal{S}_U \frac{\partial f_i}{\partial f_{i-1}}$, and so done. ∎

The dynamic programming procedure of figures 3.13-3.14 is easily adapted to compute the weak viability function. For the line:

```
g[k].psi ← g[k].psi + ψ(path);
```

is replaced by

```
g[k].psi ← 1;
```

Further, the improved dynamic programming procedure of figures 3.15-3.16 is adapted in precisely the same fashion.

4.3 The Brand-Iyengar Procedure

Brand and Iyengar have published a solution to the false path problem[11]. We have not discussed this approach in much detail yet, since we wished to have the mathematical tools in place to establish the correctness of their procedure, and to show that it gives an upper bound on the procedure developed in this paper.

The Brand-Iyengar procedure is much like the later procedure of Benkoski, et al, save that the paths are traced depth-first from the output, rather than from the inputs. A further and more important difference is the sensitization criterion. Brand and Iyengar realized the difficulty with static sensitization, referring to the errors that a straight static sensitization approach would yield as due to "a sort of circularity". Though they did not fully analyze this difficulty, they used a strategy which returns correct, though suboptimal results.

At each node f, Brand and Iyengar number the inputs to the node; for purposes of illustration, let us call these $x_1, ..., x_k$. If the current path under consideration proceeds from input x_j, then inputs $x_1, ..., x_{j-1}$ are ignored. Effectively, the Brand-Iyengar sensitizing function at node f may be written:

$$\xi^f_{x_j} = \mathcal{S}_{x_1,...,x_{j-1}} \frac{\partial f}{\partial x_j}$$

and for the path $P = \{f_0, ..., f_m\}$:

$$\xi_P = \prod_{i=1}^{m} \xi^{f_i}_{f_{i-1}}$$

We call paths P for which ξ_P is non-zero *Brand-Iyengar* paths.

Now, note that the Brand-Iyengar function is not a cover of the weak viability function. Nevertheless, we may show that this procedure is correct using theorem 4.1.1. We do this by showing:

Theorem 4.3.1 *For each path $P = \{f_0, ..., f_m\}$ weakly viable under c in a symmetric network N, there is a Brand-Iyengar path $P' = \{g_0, ..., f_m\}$ with $\xi_{P'}(c) = 1$ and $d(P) \leq d(P')$.*

Proof: Induction on $\delta(f_m)$. The base case is trivial, so assume for $\delta(f_m) < L$. Consider the case $\delta(f_m) = L$. Now, consider the set of paths weakly viable under c which terminate in an input to f_m of length at least τ_{m-1}. These paths terminate in a set of inputs to f_m, $U = \{h_0, ..., h_k\}$. Now, by induction, each such h_i terminates a Brand-Iyengar path of length $\geq \tau_{m-1}$. One of the h_i is f_{m-1}, and one is maximal under the Brand-Iyengar ordering. (The Brand-Iyengar ordering is any ordering of the inputs at all). Let h be the maximal element of U under the ordering, and V be the set of inputs to f_m which precede h in the standard order, including h. We claim that c satisfies $\xi_h^{f_m}$. Now, if $h = f_{m-1}$, then $V \supseteq U$, and

$$\xi_h^{f_m} = \xi_{f_{m-1}}^{f_m} = \mathcal{S}_{V-\{f_{m-1}\}} \frac{\partial f_m}{\partial f_{m-1}}$$

Since $V - \{f_{m-1}\} \supseteq U - \{f_{m-1}\}$, and since $\mathcal{S}_R f \supseteq \mathcal{S}_S f$ for any sets R, S such that $R \supseteq S$ and any function f, we have that:

$$\xi_{f_{m-1}}^{f_m} \supseteq \varphi_P^{f_m}.$$

and since c satisfies $\varphi_P^{f_m}$, c must satisfy $\xi_{f_{m-1}}^{f_m} = \xi_h^{f_m}$. If $f_{m-1} \neq h$, then we can do a similar calculation: the set $U - \{f_{m-1}\} \subseteq V - \{h\}$, since h is the maximal element of U under the ordering, and hence every other element of U must precede it in the ordering. Since f_m is symmetric, we have that

$$\mathcal{S}_{U-\{f_{m-1}\}} \frac{\partial f_m}{\partial f_{m-1}} = \mathcal{S}_{U-\{h\}} \frac{\partial f_m}{\partial h}$$

and since $V \supseteq U$, we therefore have

$$\mathcal{S}_{U-\{f_{m-1}\}} \frac{\partial f_m}{\partial f_{m-1}} \subseteq \mathcal{S}_{V-\{h\}} \frac{\partial f_m}{\partial h}$$

and since this function is satisfied by c, we have that $\xi_h^{f_m} = \mathcal{S}_{V-\{h\}} \frac{\partial f_m}{\partial h}$ is satisfied by c, and so in either case the claim is shown. Once the claim is given, done, since we have by the inductive assumption a Brand-Iyengar path P'' terminating in h s.t $\xi_{P''}(c) = 1$, of length $\geq \tau_{m-1}$. The path $P' = \{P'', h\}$ clearly has $\xi_{P'}(c) = 1$, and $d(P') \geq \tau_{m-1} + w(f_m) \geq d(P)$, and done. ∎

From this, we may immediately conclude from theorem 4.1.1 that the Brand-Iyengar procedure is critical-path-correct for all symmetric networks, since it is an approximation to weak viability, a known critical-path-correct criterion. This is a somewhat stronger result than Brand and Iyengar proved in their paper (their proof was only valid for gates whose values could be controlled by a single input). This result firmly establishes the correctness of their procedure. Further, Brand and Iyengar made no mention of the robustness requirement for false path elimination, and hence did not prove that their criterion was robust. This proof demonstrates that. It also, however, guarantees that the critical delay reported by the Brand-Iyengar procedure will be an upper bound on that returned by the weak viability procedure.

The bound returned by the Brand-Iyengar procedure is highly dependent on the variable ordering chosen. Consider the path in the graph that runs through the last fanin (in the standard order) to each node. The Brand-Iyengar function for this path $\{f_0, ..., f_m\}$ is

$$\prod_{i=1}^{m} \mathcal{S}_{S(f_i,P)} \frac{\partial f_i}{\partial f_{i-1}}$$

Now,

$$\mathcal{S}_{S(f_i,P)} \frac{\partial f_i}{\partial f_{i-1}} = 1$$

for every i, whence the Brand-Iyengar function for this path is 1; i.e., the path is always true by the Brand-Iyengar criterion. If the standard order is increasing in the maximum distance of a node from the primary inputs, then this path will be the longest in the graph. In other words, for every network, there is a variable order such that the longest path is true by the Brand-Iyengar criterion. This order is the worst possible for this criterion.

Better orders may be chosen; in fact, a good order is probably the reverse of the bad order for this path. This order (decreasing in distance from the primary inputs) is the one recommended by Brand and Iyengar in their paper; however, they seemed to view this as an efficiency issue rather than an issue of quality of results. They used a depth-first search with pruning, and pointed out that if an algorithm had explored the long paths first and found one true, then the search space is drastically trimmed.

In any case, the best order for this function – whatever it may be – will at best equal the weak viability criterion.

4.4 The Du-Yen-Ghanta Criteria

Other work on this phenomenon has recently emerged, most notably the work due to Du, et. al. [26]. These authors considered networks composed of simple gates, for which each input can assume either a *control* value (a value which determines the value of the gate), or a *non-control* value (broadly, an identity for the gate – e.g., 0 for OR or NOT, 1 for AND or NAND). Static sensitization can be viewed, on such networks, as asserting a non-control value on each side input of each gate along the path.

When tracing a path $P = \{f_0, ..., f_m\}$, Du et. al split the side inputs of f_i into two sets:

1. Early-arrive-signals(f_i): Those inputs g such that the length of the longest path terminating in g is $< \tau_{i-1}$; and

2. Late-arrive-signals(f_i): Those inputs g such that the length of the shortest path terminating in g is $> \tau_{i-1}$.

Note that while the intersection of these sets is always empty, their union is not necessarily equal to the set of side inputs to f_i.

These two sets may be thought of as follows: Early-arrive-signals is the set of signals that have settled to their final value before the event propagates to f_i; late-arrive-signals are those signals which undergo events only *after* τ_{i-1}.

The Du-Yen-Ghanta criterion encompasses two rules:

1. If $g \in$ Early-arrive-signals(f_i), assert a non-control value on g.

2. If Late-arrive-signals(f_i) $\neq \emptyset$, assert a *control* value on f_{i-1}

The rationale behind the first rule is by now familiar to most readers. The side inputs $g \in$ Early-arrive-signals(f_i) have already settled to their final values at τ_{i-1}, and hence must satisfy $\frac{\partial f_i}{\partial f_{i-1}}$; i.e., must have assumed non-controlling values. Note that this rule is equivalent to taking the static boolean difference and then smoothing off all the side inputs *not* in Early-arrive-signals(f_i).

It is the second rule that distinguishes this transformation. Du et al reasoned that if the path $\{f_0, ..., f_m\}$ was a longest true path, then every longer path through f_i must have been rejected as false. If $g \in$ Late-arrive-signals(f_i), then every path through g must have been rejected as false. In this case, they reasoned, a controlling value must have been asserted on the wire f_{i-1}.

Now, notice that every connection in Late-arrive-signals(f_i) must be untestable. For suppose $g \in$ Late-arrive-signals(f_i) is testable. By definition, if $g \in$ Late-arrive-signals(f_i) then no event on g can propagate to the primary outputs, otherwise there would be a longer sensitizable path running through g (since every path through g is longer than the current path). Now, if g is testable, then the value of some primary output is determined by the value of g under c; i.e, the value of g has propagated to the primary output. The value of g is also its last event, and so the last event on g has propagated to the primary output. This event must have travelled down some sensitizable path from g, contradiction. Hence g must be untestable. In a fully-testable network, therefore, only rule (1) need be considered. We call this the *weak Du* criterion.

Lemma 4.4.1 *Consider the Brand-Iyengar criterion, with the inputs to every gate ordered in decreasing order by static delay (length of longest path). Every path true by this criterion is also true by the weak Du criterion, and hence the delay estimate produced by this procedure is a lower bound on the delay estimate given by the weak Du procedure.*

Proof: Let $P = \{f_0, ..., f_m\}$ be a path true by the Brand-Iyengar criterion under this ordering. We claim that P is true by the weak Du criterion. Induction on $\delta(f_m)$. Trivial for $\delta(f_m) = 0$. Now suppose the claim holds for $\delta(f_m) < N$. If $\delta(f_m) = N$, by induction the statement holds for $\{f_0, ..., f_{m-1}\}$. We must show that it holds for $\{f_0, ..., f_m\}$. Let f_{m-1} be the kth input of n to f_m under the ordering. Now, each input of order $< k$ terminates a path of length $\geq \tau_{m-1}$ by the definition of the order, no such input is in Early-arrive-signals(f_m). Hence under the weak Du criterion each input in the range $1, ..., k$ is left unspecified, i.e., smoothed off. Hence the set of signals left unspecified by the weak Du criterion is a superset of those left unspecified by Brand-Iyengar under this order; i.e., the path logic function corresponding to the weak Du criterion contains the Brand-Iyengar function under this order, giving the result. ∎

Note that lemma 4.4.1 tells us that Brand-Iyengar is a lower bound on the strong Du criterion (that using both rules) only on fully-testable networks. On networks with redundant connections, the delay estimate given by the strong Du criterion may be unequal to that given by the weak Du criterion. In any case, it may be shown that the strong Du criterion is weaker than viability; this is shown in the following theorem, which demonstrates that, if there is a longest viable path $P = \{f_0, ..., f_m\}$, viable under c such that for node f_i Late-arrive-signals(f_i) is non-empty, then c sets f_{i-1} to a controlling value. This suffices to show that the strong Du criterion is an approximation to the viability criterion, and hence the strong Du criterion is a correct, robust criterion by the approximation theorem.

Theorem 4.4.1 *Let N be a symmetric network. Let $P = \{f_0, ..., f_m\}$ be a longest viable path through f_i, P viable under c. Let g be a side input to f_i such that every path through g is longer than τ_{i-1}. Then either f_{i-1} is set to value a by c such that*

$$\left.\frac{\partial f_i}{\partial g}\right|_{f_{i-1}=a} \equiv 0$$

or there is another viable path under c at least as long as P.

Note that if simple gates are assumed, this simplifies to the statement that f_{i-1} is set to a controlling value for f_i by c.

Proof: Assume that

$$\left.\frac{\partial f_i}{\partial g}\right|_{f_{i-1}=a} \not\equiv 0$$

Then we claim that for each $i \leq k \leq m$ we can construct a path P_k at least as long as P, viable under c, terminating in f_k. Induction on k. For $k = i$, let U be the set of inputs to f_i terminating viable paths under c of length $\geq \tau_{i-1}$. Now, consider the set of paths viable under c terminating in an input to f_i of length $\geq \tau_{i-1}$, excluding the path $\{f_0, ..., f_{i-1}\}$. Note this set is nonempty since the set of shortest paths through g are all viable under 1, and are of length $> \tau_{i-1}$. Since the set is nonempty, it has an element of least length, P_h, and P_h terminates in h. We claim that $\{P_h, f_i\}$ is viable under c. Consider the term of the viability series for P_h:

$$\mathcal{S}_{U-\{h\}} \frac{\partial f_i}{\partial h} \prod_{h' \in U-\{h\}} \psi^{h',|P_h|}$$

Since P_h is chosen to be minimal in the set, every member of U aside from h terminates a viable path of length $\geq P_h$, i.e. $\psi^{h',|P_h|}$ is satisfied by c for every h' in $U - \{h\}$. Now, all we must show is that

$$\mathcal{S}_{U-\{h\}} \frac{\partial f_i}{\partial h}$$

is satisfied by c. Since f_i is symmetric, we have three cases:

1.
$$\mathcal{S}_{U-\{h\}} \frac{\partial f_i}{\partial h}$$
   is satisfied by c; or

2.
$$\mathcal{S}_{U} \frac{\partial f_i}{\partial f_{i-1}}$$
   is not satisfied by c; or

3. f_{i-1} is set to a value a by c such that
$$\mathcal{S}_{U-\{h\}} \frac{\partial f_i}{\partial h}\bigg|_{f_{i-1}=a} \equiv 0$$

Case 1 proves the claim, so we must dispose of (2) and (3). (2) is false, since if it were true, P would not be viable under c, contradiction. For case 3, since $\mathcal{S}_V f \equiv 0 \Rightarrow f \equiv 0$, we have that:

$$\frac{\partial f_i}{\partial h}\bigg|_{f_{i-1}=a} \equiv 0$$

But then, since f_i is symmetric:

$$\frac{\partial f_i}{\partial g}\bigg|_{f_{i-1}=a} \equiv 0$$

Contradiction. We are left with case (1), which proves the claim: $\{P_h, f_i\}$ is viable under c. Hence the statement of the theorem holds for the case $k = i$; the path $\{P_h, f_i\}$ is viable under c, is of length $\geq \tau_i$ and is not equal to $\{f_0, ..., f_i\}$. Assume inductively that the statement holds for $k < K$. Consider the case $k = K$. By induction, we have a path P_{K-1} terminating in f_{K-1} viable under c at least as long as $\{f_0, ..., f_{K-1}\}$ that is *not* $\{f_0, ..., f_{K-1}\}$. Now, we have two cases:

1. f_K is statically sensitized to f_{K-1} by c, in which case the claim holds for f_K;

2. There are a set of paths U_K at least as long as $\{f_0, ..., f_{K-1}\}$ terminating in some side inputs to f_K. Of the set of paths $U_K \cup \{\{f_0, ..., f_{K-1}\}\}$ at least one had a viable extension under c through f_K. Hence the set of paths $U_K \cup \{P_{K-1}\}$ met the assumptions of lemma 2.4.2, and so by the terms of that lemma at least one of these paths has a viable extension under c through f_K. This path is not $\{f_0, ..., f_K\}$ and is of length $\geq \tau_K$, proving the theorem.

■

This theorem suffices to demonstrate that the strong Du criterion is an approximation to the viability criterion, since it demonstrates that the use of Du's second rule does not affect the viability of the longest viable path. Further, in practice, only a few connections will be untestable, and the second rule need not apply to all of those. Hence the second rule will be invoked only rarely. Hence, the strong Du and weak Du algorithms should only rarely give different bounds.

4.5 The Chen-Du Criterion

More recently, H. C. Chen and D. H. C. Du at the University of Minnesota have presented a new criterion[21], which can be regarded as an incremental improvement on viability on networks composed of simple gates. This criterion, as we shall see below, returns the same delay bound as the viability criterion on these networks, though possibly more conservative bounds on networks consisting of more complex gates. However, the Chen-Du criterion does represent an advance on the state of the art. Chen and Du were concerned with the following problem. Given a network N, and a time τ, identify the minimal set of paths S such that, if every $P \in S$ is reduced in delay so that $d(P) \leq \tau$, then the delay of the longest true path in the network is $\leq \tau$. Chen and Du demonstrated that the viability algorithm returned a non-minimal set of such paths; further, they demonstrated that the dynamic sensitization criterion returned a non-minimal set of such paths. As a result, their algorithm and criterion is of interest from the point of view of performance enhancement.

As with the Du-Yen-Ghanta criterion, the Chen-Du criterion is concerned with networks composed of simple gates. We can then talk about "controlling" values for gates, and "non-controlling" values for gates. Recall that the boolean difference formulation was simply the extension to complex gates of the idea that each side input have a non-controlling value; on simple gates the two ideas are identical.

The Chen-Du criterion differs from viability in that the nature of the on-path signal is considered. The authors reasoned as follows. If a gate has non-controlling values on all of its inputs, then the *last* input to stabilize controls the delay of the gate. However, if a gate has controlling values on some of its inputs, then the *first* signal that stabilizes with a controlling value controls the delay of the gate. Chen and Du are only interested in paths that control the delay of every gate along the path; as a result, their criterion for a path $P = \{f_0, f_1, ..., f_m\}$ to be *true* is that for each i, f_i controls the delay for f_{i+1}. The condition under which each f_i controls the delay for f_{i+1} breaks into two cases. If f_i has a control value, then each side input to f_{i+1} either has a non-control value or stabilizes no earlier than τ_i. If f_i has a non-control value for f_{i+1}, then each side input to f_{i+1} must both have a non-control value for f_{i+1} and stabilize no earlier than τ_i. Let us denote the condition that signal f has a control value for g as $c(f, g)$ and the condition that f has a non-control value as $nc(f, g)$, Further, let us denote as ω_P the condition that the Chen-Du criterion is met for path P. We can then write the Chen-Du function relatively easily.

As before, we write the Chen-Du function for path $P = \{f_0, ..., f_m\}$, as a product of functions at each node

$$\omega_P = \prod_{i=1}^{m} \omega^{f_i, P} \tag{4.4}$$

Now, the condition that a node g stabilizes later than t is that there is a Chen-Du path of length $> t$ terminating in g. Denoting the set of paths terminating in g of length $> t$ as $\mathcal{Q}_{g,t}$, we can write the condition "g stabilizes later than t" as $\omega_1^{g,t}$:

$$\omega_1^{g,t} = \sum_{Q \in \mathcal{Q}_{g,t}} \omega_Q \tag{4.5}$$

By a similar argument, if we denote the set of paths of length *exactly* t terminating in g as $\mathcal{R}_{g,t}$, we can write the condition "g stabilizes at exactly t" as $\omega_2^{g,t}$:

$$\omega_2^{g,t} = \sum_{Q \in \mathcal{R}_{g,t}} \omega_Q \tag{4.6}$$

We are now in a position to write equations corresponding to the two cases for the Chen-Du sensitization criterion. We have that if f_i is at a controlling value for f_{i+1}, then each side input g must either be at a non-controlling value for f_{i+1} or stabilize no earlier than τ_i. The condition that g stabilize no earlier than τ_i means that g must stabilize either later than τ_i or at exactly τ_i. We therefore have:

$$c(f_i, f_{i+1}) \prod_{g \in S(f_{i+1}, P)} (nc(g) + \omega_1^{g,\tau_i} + \omega_2^{g,\tau_i})$$

Now, if f_i is at a non-controlling value for f_{i+1}, we must have that each side input g:

1. Stabilizes no later than τ_i. Since g must stabilize at some time, this condition is therefore the complement of the condition that g stabilizes later than τ_i, i.e., this condition is $\overline{\omega_1^{g,\tau_i}}$; and

2. g is at a non-control value for f_{i+1}, i.e., $nc(g, f_{i+1})$.

We therefore have:

$$nc(f_i, f_{i+1}) \prod_{g \in S(f_{i+1}, P)} (nc(g, f_{i+1})\overline{\omega_1^{g,\tau_i}})$$

Putting the cases together, we have:

$$\omega_P^{f_{i+1}} = \begin{array}{l} c(f_i, f_{i+1}) \prod_{g \in S(f_{i+1}, P)} (nc(g, f_{i+1}) + \omega_1^{g,\tau_i} + \omega_2^{g,\tau_i}) + \\ nc(f_i, f_{i+1}) \prod_{g \in S(f_{i+1}, P)} (nc(g, f_{i+1})\overline{\omega_1^{g,\tau_i}}) \end{array} \tag{4.7}$$

Two theorems are required to affix this criterion firmly in our spectrum.

Theorem 4.5.1 *For each path $P = \{f_0, ..., f_m\}$ in a network composed of simple gates, $\psi_P \supseteq \omega_P$*

Proof: Induction on $\delta(f_m)$. For f_m a primary input, trivial, so assume for $\delta(f_m) < N$. If $\delta(f_m) = N$, all we must consider is the relationship between $\psi_P^{f_m}$ and $\omega_P^{f_m}$. Now, since f_m is a simple gate, $\psi_P^{f_m}$ can be written:

$$\prod_{g \in S(f_m, P)} (nc(g, f_m) + \psi^{g,\tau_{m-1}})$$

Noting that:

$$(nc(g, f_{i+1})\overline{\omega_1^{g,\tau_i}}) \subseteq (nc(g, f_{i+1}) + \omega_1^{g,\tau_i} + \omega_2^{g,\tau_i})$$

We can write that

$$\omega^{f_m,P} \subseteq (nc(f_{m-1}, f_m)+c(f_{m-1}, f_m)) \prod_{g \in S_{f_m,P}} (nc(g, f_m)+\omega_1^{g,\tau_{m-1}}+\omega_2^{g,\tau_{m-1}})$$

Note that $(nc(f_{m-1}, f_m) + c(f_{m-1}, f_m)) = 1$ and that, by induction:

$$\omega_1^{g,\tau_{m-1}} + \omega_2^{g,\tau_{m-1}} \subseteq \psi^{g,\tau_{m-1}}$$

Simple substitution then yields the result. ∎

By the approximation theorem, then, one can say immediately that the Chen-Du criterion yields a lower bound on the delay reported by viability: every Chen-Du path is also a viable path, but the converse is not necessarily true. In order to prove that the Chen-Du criterion also yields an *upper* bound on the delay reported by viability, we must use the *second* clause of the approximation theorem: that is, for each viable path, we must construct a Chen-Du path at least as long.

Theorem 4.5.2 *For each path $P = \{f_0, ..., f_m\}$ in a network composed of simple gates, such that $\psi_P \supseteq c$, there is a path $Q = \{g_0, ..., g_n, f_m\}$ such that $d(Q) \geq d(P)$ and $\omega_Q \supseteq c$*

Proof: Let $P = \{f_0, ..., f_m\}$ be the longest viable path under c terminating in f_m. Induction on $\delta(f_m)$. For f_m a primary input, trivial, so assume for $\delta(f_m) < N$. If $\delta(f_m) = N$, then we have the case for each fanin g of f_m. We have two cases.

1. $c \in c(f_{m-1}, f_m)$. Since $\psi_P^{f_{m-1}} \supseteq c$ by assumption, we must have for each side input g $c \in nc(g, f_m)$ or $c \in \psi^{g,\tau_{m-1}}$. If $c \in \psi^{g,\tau_{m-1}}$, we must have that $c \in \omega_Q$ for some Q terminating in g of length $\geq \tau_{m-1}$, and hence by induction $c \in \omega_1^{g,\tau_{m-1}} + \omega_2^{g,\tau_{m-1}}$. Hence $c \in \omega_P^{f_m}$.

2. $c \in nc(f_{m-1}, f_m)$. Let U be the set of fanins g such that $c \in c(g, f_m)$. If $U = \emptyset$ (i.e., each fanin of f_m is set to a non-control value by c) then let R be a longest path terminating in some fanin g to f_m such that $\psi_R \supseteq c$. By induction, we know that there is some path R_1 at least as long terminating in g such that $\omega_{R_1} \supseteq c$.

Now, let R_2 be the longest Chen-Du path under c terminating in a fanin to f_m. By construction, $d(\{R_2, f_m\}) \geq d(\{R_1, f_m\}) \geq d(P)$. Further, since R_2 is a longest Chen-Du path under c terminating in a fanin to f_m, then for each fanin g to f_m, $c \subseteq \overline{\omega_1^{g,d(R_2)}}$. Hence $c \in nc(g) \subseteq \overline{\omega_1^{g,d(R_2)}}$ for each fanin g of f_m, hence $\{R_2, f_m\}$ is a Chen-Du path under c and is at least as long as P. Now, if $U \neq \emptyset$, for each $g \in U$ let R_g be the longest viable path under c terminating in R_g. Note that each R_g is of length $\geq \tau_{m-1}$, for we must have for each $g \in S(f_m, P)$, c satisfies

$$nc(g, f_m) + \psi^{g,\tau_{m-1}}$$

and we know that c does not satisfy $nc(g, f_m)$ for all $g \in U$ Now, by induction, for each R_g there exists a P_g terminating in g of length $\geq d(R_g)$. Let P_h terminating in h be of minimal length among the P_g. We claim that $\omega_{P_h}^{f_m} \supseteq c$. All we must show is that for every $g \in$fanins(f_m), either $c \in nc(g, f_m)$ or $c \in \omega_1^{g,d(P_h)} + \omega_2^{g,d(P_h)}$. Now, if $c \notin nc(g, f_m)$ then $g \in U$, and by construction for each such g there is a path P_g terminating in g of length $\geq d(P_h)$ such that $\omega_{P_g} \supseteq c$. Noting that:

$$\omega_1^{g,d(P_h)} + \omega_2^{g,d(P_h)}$$

is simply the sum of the Chen-Du functions of paths terminating in g of length $\geq d(P_h)$, we have that

$$c \subseteq \omega_1^{g,d(P_h)} + \omega_2^{g,d(P_h)}$$

for each $g \in U$. Hence

$$c \subseteq nc(g, f_m) + \omega_1^{g,d(P_h)} + \omega_2^{g,d(P_h)}$$

for each g in the fanins of f_m. By construction, therefore, $\omega_{P_h}^{f_m} \supseteq c$, hence $\{P_h, f_m\}$ is a Chen-Du path under c, and is at least as long as P.

■

These two theorems lead us to the final conclusion:

Corollary 4.5.3 *The Chen-Du criterion and the strong viability criterion concur on the delay of the longest sensitizable path on simple networks.*

Note that this does not say that Chen-Du and viability will always give the same answer. Chen-Du derives its power from being a criterion defined on networks of simple gates. Viability, in contrast, is a criterion defined on networks of complex gates and is proved robust on networks consisting of symmetric gates. Complex gates are handled by Chen-Du in the usual fashion, by macroexpansion. However, this macroexpansion increases the set of Chen-Du paths and so potentially increases the delay estimate. Hence, viability can give more aggressive estimates than Chen-Du if symmetric, complex gates are present in the network.

At this writing, it is strongly conjectured that the set of paths returned by the Chen-Du criterion on networks of simple gates is the same as that returned by the improved dynamic programming procedure on those networks.

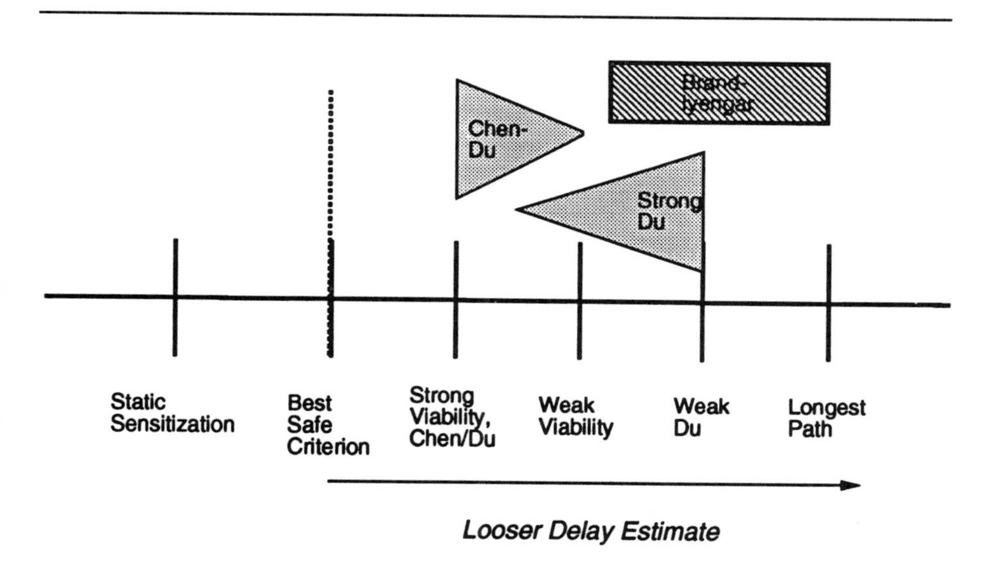

Figure 4.1: Sensitization Criteria

This ends our discussion of sensitization criteria that appear in the literature. The relative tightness of the various criteria are shown graphically in figure 4.1. In this diagram, "tightness" increases from left to right. Simple longest path marks the far right edge. "Best Safe Criterion" marks the boundary between safe (that is, correct and robust) criteria and unsafe, overly aggressive criteria. The rectangle for Brand-Iyengar represents a spectrum of criteria, parameterized by order. The triangles for the strong Du and the Chen-Du criteria represents the fact that the tightness of these criteria as opposed to others depends upon the exact circuit. The height of the triangle at any given point on the x axis denotes how probable the answer given by such a variable criterion is to return the answer given by the criterion which corresponds to that point on the x axis; as a result, Chen-Du will give the same answer as viability on almost every circuit (the exceptions being the circuits containing symmetric, complex gates), and strong Du will give the same answer as weak Du on almost every circuit. The Brand-Iyengar algorithm can give better or worse answers than the strong Du criterion: it is better on some circuits, worse on others, but is most often equal to the weak Du criterion on circuits; it is always bounded by the range given. As can be seen from the diagram, viability and Chen-Du share the honour of being the tightest known safe criteria.

4.6 More Macroexpansion Transformations

Theorem 2.5.1 is valid for *any* macroexpansion transformation. We have only used it for the *and/or* transform, since it appears that this introduces the fewest *spurious* viable paths, and hence yields the lowest bound on the critical delay. However, it is possible that other transforms, though yielding poorer results, might well entail much less expensive computation. In particular, consider the *two-input nand* transform.

It is well-known that any function can be realized by a network of two-input nand gates. The two-input nand transform of a node f is any transform of f meeting the conditions of definition 2.5.1. This transform is attractive when one considers that there is at most one side input to a gate on any path $P = \{f_0, ..., f_m\}$ in a two-input network, and (if we represent the side input to node f_i as g_i), then the only subsets of the set of side inputs to f_i are $\emptyset$ and g_i. Hence g_i is the only possible input to $\frac{\partial f_i}{\partial f_{i-1}}$, and thus $\mathcal{S}_{g_i}\frac{\partial f_i}{\partial f_{i-1}} = 1$. Given this, we can simplify the definition of the viability function for a path in such a network as:

$$\psi_P = \prod_{i=0}^{m} \psi_P^{f_i}$$

$$\psi_P^{f_i} = \frac{\partial f_i}{\partial f_{i-1}} + \psi^{g_i,\tau_{i-1}}$$
$$\psi^{g,t} = \sum_{Q \in \mathcal{P}_{g,t}} \psi_Q$$

This function is much simpler to compute than the viability function of definition 2.6.2, since the power-set sum in $\psi_P^{f_i}$ has disappeared. Moreover, the network is symmetric, and hence its correctness is guaranteed by the theorems previously proved. The reader is cautioned, however, that an increase in the number of paths may be expected in such a network, in which case some percentage of the gains realized may be offset.

This observation suggests that the algorithm presented above may be considered a *family* of algorithms, with different performance and result characteristics. The members of the family may be distinguished by the transform taken.

4.7 Biased Satisfiability Tests

Given that the algorithms enumerated so far all have a satisfiability test in the inner loop, it is unsurprising that these procedures take a very long time. Nevertheless, one may hope to use the approximation theorem to yield up a hint of polynomial-time approximations in this domain as well. Consider any positively-biased satisfiability test; i.e., a procedure which guarantees to report that a function f is nonzero when it is not identically zero, but may occasionally report that a function is nonzero when it is identically zero. Such a procedure may be considered an *exact* satisfiability test on a function $\hat{f} \supseteq f$. This consideration yields immediately the conclusion that if γ is any critical-path correct path logic function, then a procedure using γ as the sensitization criterion with a positively biased satisfiability test will also yield a critical path correct family of logic functions.

There are undoubtedly many such biased tests. We detail one here, and more fully in appendix C. Conceptually, one can think of such a test on a multi-level network as follows.

The functions we have enumerated above are, in general, functions of not only the primary inputs but also of the intermediate nodes in the network. From a theoretical point of view, this is a distinction without a difference; since the intermediate nodes themselves realize functions of the primary inputs, the functions we have been describing above must necessarily be functions of the primary inputs; moreover, that function may be (conceptually) realized by substituting the function (in terms of the primary inputs) for each intermediate node. This process is known as *collapsing.*

Mechanically, the way that functions are discovered to be 0 in the collapsing process is that for each cube of f, there is some x such that both x and $\overline{x}$ appear in the cube; i.e., for the cube to be satisfied we must have $x = 1$ and $x = 0$. Since this is clearly not possible, the cube is unsatisfiable. More generally, each such cube of the function is found to be inconsistent; no primary input vector gives rise to such a cube.

This gives us a clue as to how to proceed. Consider a procedure that directly simulates the effect of attaching values to the various wires in the circuit. Such a simulation will not be entirely complete, in the sense that the effect of the assertions will not be carried back to the primary inputs. Under such a simulation, a function is determined to be 0 if for each cube there is some x such that both x and $\overline{x}$ appear in the cube; i.e., each cube is *explicitly* inconsistent. Now, it is obvious that every cube that is explicitly inconsistent is inconsistent; however, the converse is not the case. Hence this is a positively-biased satisfiability test.

Such simulations appear to be done by Benkoski et al [6]; he uses a "D-Algorithm without justification", which presumably means computing implications. Brand and Iyengar also compute implications, and refer specifically to the NOR-gate rules formulated by Trevillyan, Berman, and Joyner[8]. We have reformulated those rules for general gates, and present an efficient algorithm for computing these in appendix C; this algorithm features a quartic preprocessing phase and a quadratic main phase. The difference in order is important, for only the quadratic main phase is called during path tracing.

4.8 Axes of Approximation

The theory developed above leads us to the conclusion that there are several dimensions of approximation, more or less orthogonal. Along one axis lies the sensitization criterion; this we have already detailed. Along a second lies the satisfiability test. Along a third lies the macroexpansion transform. Because few other procedures use macroexpansions, we omit them in this picture.

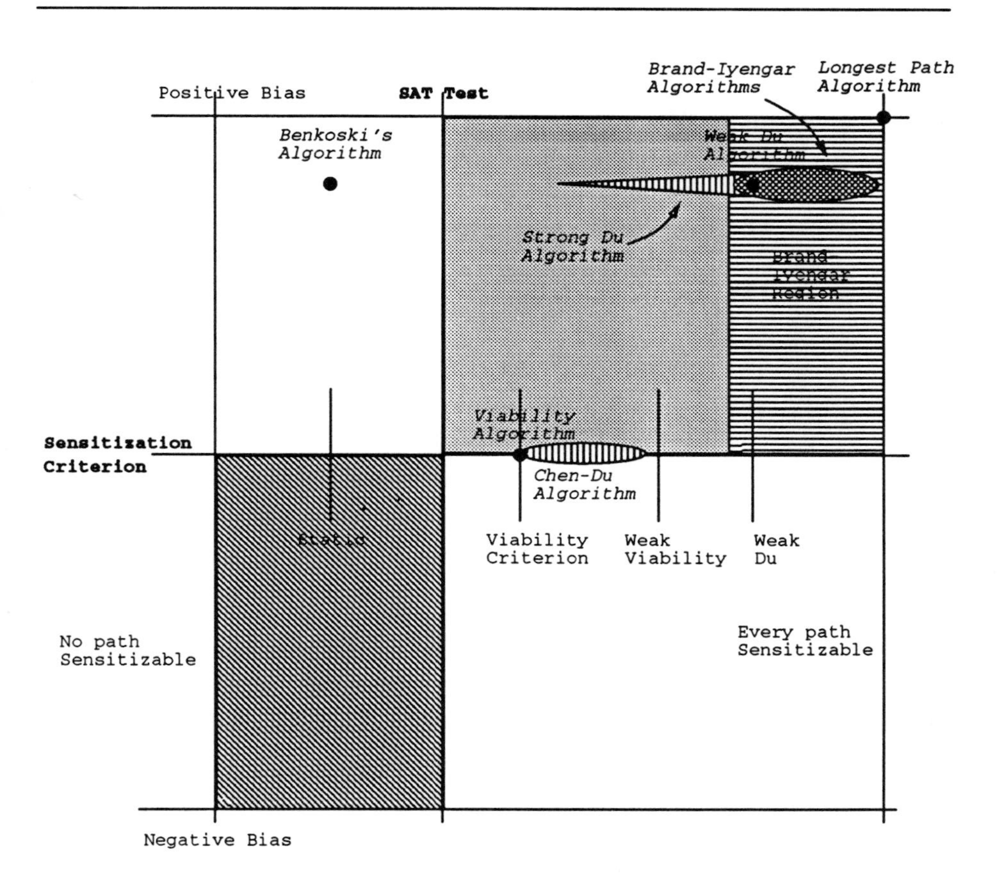

Figure 4.2: Axes of Approximation

The two major axes of approximation – sensitization criterion and satisfiability test – are depicted in figure 4.2. The x dimension represents sensitization criterion, and ranges from too restrictive (no paths sensitizable) to too loose (every path sensitizable). Similarly, the SAT test axis ranges from negative bias (all functions reported unsatisfiable) to positive bias (all functions reported satisfiable). Both dimensions increase in the direction of safety. The origin of the graph is the exact, minimum criterion. Note that no known sensitization criterion is at the origin.

Hence, programs depicted in the upper-right quadrant are known safe; they will always return an upper bound on the true delay. Programs in the lower-left quadrant are known unsafe; they will always return a lower bound on the true delay. Programs in the other quadrants give no guarantees at all.

4.9 The Llama Timing Environment

The observation of section 2 that the false path detection procedures were all parameterized variants of the same algorithm led to the development of a program which serves as an experimental testbed for these procedures. This system, the LLAMA timing environment, has been built on top of the MIS II logic synthesis system at UC-Berkeley [15, 13], and uses the underlying facilities of MIS for boolean function manipulation and for the extraction of delay information at the nodes. Further, the MIS II command interpreter is used to permit user-level experimentation with the parameters of LLAMA.

LLAMA is parameterized on five distinct axes: search method (best-first or depth-first), sensitization criteria (static, viability, or Brand-Iyengar), satisfiability test (test to determine whether the sensitization function is 0), representation of functions (Bryant's graph-based representation[19][58] vs sum-of-products form), and delay model. The selection of the various parameters is made at the beginning of a timing run by the user through MIS II command-line switches.

To date, we have experimented extensively with function representation and the various sensitization criteria. We have used two delay models in the calculations, a *unit* delay model, under which each gate has unit delay, and a *library* delay model. Under the latter, the network has been mapped to a network of standard cells, each of which has a

Ckt	O/M	Long	Brand	Viable	Static
5xp1	OM	21.80	21.80	20.40	19.20
5xp1	M	19.80	19.80	18.40	18.40
C7552		43.00	43.00	42.00	42.00
Des	O	11.00	11.00	10.00	10.00
Des	OM	68.20	68.20	66.40	64.00
Rot	O	10.00	10.00	9.00	9.00
Rot	OM	29.60	28.60	27.20	27.20

Table 4.1: Critical Delay of Benchmark Circuits

well-characterized delay.

Before the main loop of the algorithm is entered, LLLAMA goes through a pre-processing phase. During this phase, a static delay trace is done to compute the delays and the esperance of each node, the static sensitization and Brand sensitization functions for each input of each gate are computed, and the variables `g[k].psi` are set to 0 for each input `k` to gate `g`. Further, asymmetric gates are macroexpanded into subnetworks of symmetric gates, if necessary. The data structure of partial paths is initialized to the set of primary inputs of the circuit.

4.10 Experimental Results

LLLAMA has been run on two broad classes of circuits: the public benchmark circuits, and parameterized circuits which are known to contain false paths. For each circuit, we report the critical delay according to the longest path procedure and the static, Brand-Iyengar, and viability conditions. We also report whether the circuit was optimized by the MIS-II standard script (O), and whether the circuit was mapped to the MSU standard-cell library (M). Mapped circuits have delays reported by the mapped model; unmapped circuits have delays reported by the unit model.

Two sets of public benchmark circuits were run: the IWLS and ISCAS benchmark suites. Of the IWLS circuits, only the benchmark circuit **5xp1** showed any false paths under any criterion. The benchmarks DES and ROT were also run. Though not part of the IWLS benchmark suite, these circuits are available from UC-Berkeley in BLIF format. Of the ISCAS circuits, C880, C432, C499, and C17 all had no false paths. C7552 is

Bits	Block	O/M	Long	Brand	Viable	Static
8	2		13.00	13.00	8.00	8.00
8	2	M	17.80	15.40	15.40	15.40
8	4		11.00	11.00	10.00	10.00
8	4	M	14.80	13.60	13.60	13.60
16	2		25.00	25.00	12.00	12.00
16	2	M	35.40	27.40	27.40	27.40
16	4		21.00	21.00	12.00	12.00
16	4	M	29.20	20.00	20.00	20.00
32	4		41.00	41.00	16.00	16.00
32	4	M	58.00	32.00	32.00	32.00

Table 4.2: Critical Delay of Carry-Bypass Adders

shown in the table. The remainder not shown failed to complete.

It has long been known that carry-bypass adders exhibit false paths.[1] A final set of experiments involved the generation of carry-bypass adders of varying sizes, and block sizes for the bypass chain. Integers N and M in the first two columns of table 4.2 represents an N-bit adder, with M bits in the bypass chain. BLIF descriptions of these circuits, and a program to generate the BLIF description from arbitrary N and M, are available from the authors.

[1] Prof. H. De Man kindly brought this fact to the attention of the authors

Chapter 5

Hazard Prevention in Combinational Circuits

5.1 Introduction

Previous research into timing properties of circuits has led to considering the problem of *hazards* or *glitches* in combinational circuits. One can demonstrate that in the absence of hazards, a variety of strong properties hold which are not valid in the general case: in particular, in the next chapter we will show that timing analysis can obtain tight bounds on the critical path of a circuit, which was shown to be impossible for a hazardous circuit.

Given these desirable properties, it is worth considering whether certifiably hazard-free circuits may be synthesized. It is well-known that the class of *precharged unate* circuits (e.g., NORA, DOMINO, and DCVS) are hazard-free; indeed, these circuits can only function if they are hazard-free. The characteristics of these circuits are reviewed in appendix D. What we wish to discover is whether any fully-restoring circuits are hazard free.

The remainder of this chapter is organized as follows. In section 5.2, hazards will be defined and a set of assumptions concerning the way that signals change. In section 5.3 we relate function evaluation with walks on the Boolean n-cube, and show that every hazard-free circuit is precharged-unate, under our assumptions. In sections 5.4-5.5, we relax two of the initial assumptions and show that the results of section 5.3 hold even if either, or both, of these assumptions are relaxed.

5.2 Hazards

A *hazard* at a node is a multiple change in its value during an evaluation period (typically, say, a clock cycle or phase). Statically, we view a logic functions as an ideal switch, which remains at a value until some input has switched and then immediately switches to the new, output value. Further, in analyzing hazard-free circuits, we make the following assumptions about the behaviour of the inputs to the node:

1. The inputs to a node begin in some initial steady state and change, one at a time, until they reach some final steady state.
2. Each input to a node may change at most once during the evaluation of a node; this condition is assured if every node in, and every primary input of, the circuit is hazard-free
3. Each input to a node may change during evaluation.
4. The order in which the inputs change is unpredictable.
5. No combination of the inputs is forbidden as either the initial nor the final steady state.

These assumptions are strong in the sense that they enchance hazards. All but the last two are easily justified. The last is not only unjustifiable, it is generally false. In practice, only a few states can occur. Indeed, multi-level logic optimization makes heavy use of such forbidden states, which are vectors covered by the satisfiability don't-care set[3]. We will be relaxing this assumption later.

The fourth assumption is also a little shaky. In practice, some rough guesses can be made, though, as we shall see below, due to the statistical nature of delays in a MOS circuit, precise orders on variable arrival can only be guaranteed at some cost in circuit speed. However, as will be demonstrated, neither redundancy information nor information on variable order affect the major results of this paper.

Eichelberger [27] attempted to characterize hazard-free circuits, and detect hazards. He concluded that *function hazards* were unremovable, whereas *M-hazards* were removable by adding redundancy to a two-level realization of a circuit. Function hazards are those inherent in the cube representation of the function, and occur in every non-trivial Boolean function. *M-hazards* are those that occur due to differences in arrival

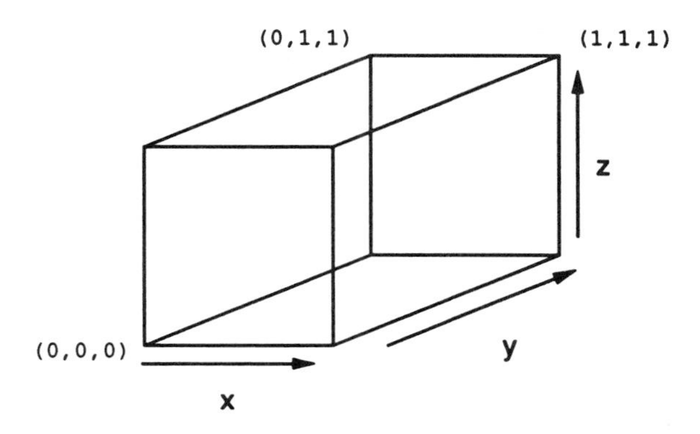

Figure 5.1: The Boolean 3 Cube

times of wires attached to the same net (so, for example, a variable y may show a 1 on one lead of the net and a 0 on another lead).

Breuer and Harrison [18] attempted to design tests that would not excite hazards of a circuit. They designed a multivalue calculus to detect necessary and sufficient conditions for tests to not excite hazards in a circuit. Their results for hazard-free circuits required that all the gates in the network be unate, and that the initial input vector on each gate be $\vec{0}$ or $\vec{1}$.

5.3 The Boolean n–Space

In general, an n input boolean function can be described as a set of points in n-space. Since each coordinate of each point in the space is either 0 or 1, it is natural to view each point as a vertex of an n-dimensional cube. Each vertex of the cube represents a unique combination of input values, and hence there is a value of the function at that point. In the diagrams in this paper, the vertices where the function is 1 are shaded. The initial values of the inputs form one vertex of the cube; the final values of the inputs are at the furthest vertex from the initial vertex[1]. The initial value of the function is its value at the initial vertex; As the inputs change (assuming each input undergoes exactly one change), we move to a corresponding vertex on the cube, terminating finally at the

[1]We can assume that if any variable does not change, then it is not a dimension of the n-cube; hence we can assume that all variables change value

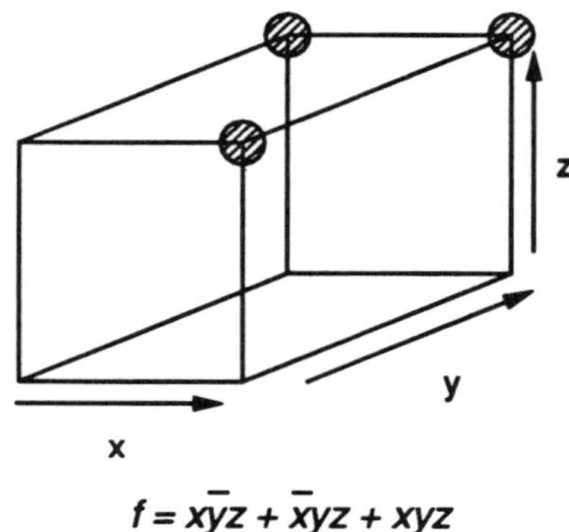

Figure 5.2: A Function on the Boolean 3 Cube

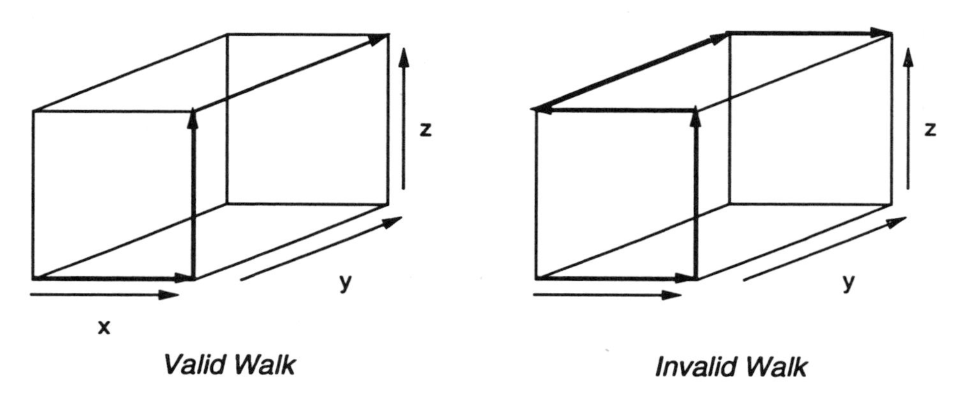

Figure 5.3: Valid and Invalid Walks on the N-Cube

final vertex. Since the inputs change one at a time, we move a distance one along the cube for each change; since each input changes exactly once, we make precisely n moves. (If an input does not change, we may consider that there are only $n-1$ variables for the purposes of this discussion).

As we make each move, the value of the function changes to the value of the visited vertex. A *hazard* exists iff the value of the function changes more than once. We formalize these intuitive notions as follows.

Definition 5.3.1 *A sequence (or "walk") of vertices* $v_0, ..., v_j$ *is said to be* **valid** *iff* $dist(v_i, v_{i+1}) = 1 \; \forall i$ *and* $dist(v_0, v_j) = j$. *If* $j = n$, *the sequence is said to be a full walk, and* v_n *is said to be* v_0*'s* **off-vertex**, *denoted* $\overline{v_0}$

Given the assumptions under which we are working, we can make some observations about the number of different walks on the $n-$ cube. At distance exactly k from v_0 there are $2^{\min(k,n-k)}$ unique vertices. Now, on a walk of length k, k variables change, and since they can change in any order they give rise to $k!$ distinct sequences. Hence there are precisely $2^{\min(k,n-k)}k!$ sequences of length k, and so

$$\sum_{k=0}^{n} 2^{\min(k,n-k)}k!$$

valid sequences beginning at some vertex v_0. We are particularly interested in the set of length n sequences. There are $2^n n!$ such sequences, and each such sequence terminates at the *unique* vertex $\bar{v_0}$.

We consider valid walks and their properties. Assumptions 1-5 guarantee that every evaluation of a function corresponds to some valid sequence. The value of the function will change at least twice during the walk iff the walk corresponds to a hazard. We can then characterize walks in terms of the number of times that a function changes value.

Definition 5.3.2 *A valid sequence of vertices* $\{v_0, ..., v_n\}$ *is said to be* **monotone** *iff, for every* $0 \leq i \leq n$, $f(v_i) = f(v_0)$ *implies* $f(v_j) = f(v_0)\ \forall j \leq i$.

Non-monotone walks and hazards are equivalent. We can immediately say, when assumptions 1-5 hold:

Theorem 5.3.1 *Let* v *be any vertex on the n-cube,* f *any function with its inputs hazard free.* f *undergoes a hazard during the transition from initial input state* v *to final input state* $\overline{v}$ *iff there is at least one non-monotone walk from* v *to* $\overline{v}$

Proof: Let f undergo a hazard. Now, the transition from v to $\overline{v}$ for the inputs involves some walk on the n-cube, $\{v, v_1, ..., v_{n-1}, \overline{v}\}$, and f assumes the value $f(v_i)$ as the walk transits through v_i. Since f undergoes a hazard, there is some least j such that $f(v_{j-1}) = f(v)$, $f(v_j) \neq f(v_{j-1})$ (the first transition of f), and there is some least $k > j$, such that $f(v_j) = f(v_{j+1}) = ... = f(v_{k-1}) \neq f(v_k)$ (the second transition, inducing the hazard). Such a walk is non-monotone. Conversely, suppose

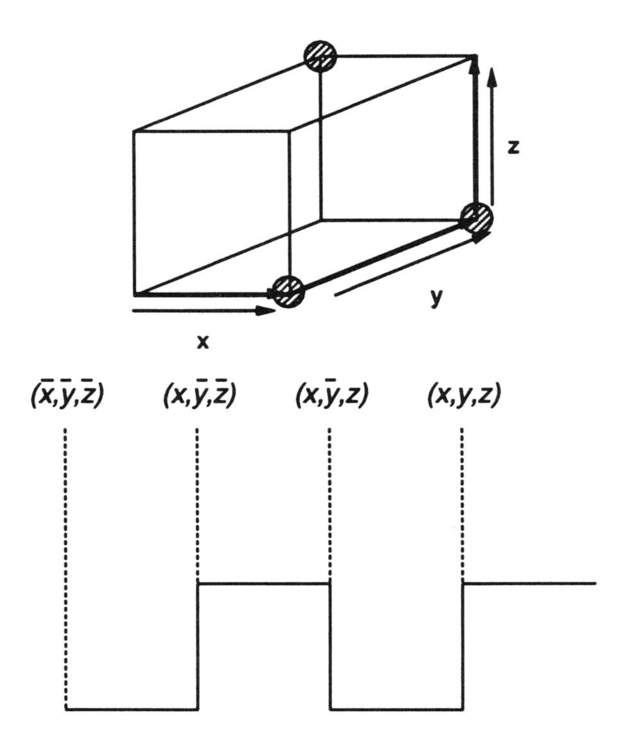

Figure 5.4: Hazards Arising from a Walk

there exists a non-monotone walk. To each valid walk there corresponds an order on the arrival time of the variables, and we have assumed that any variable order may occur, so choose the variable order that corresponds to the non-monotone walk. This order induces the non-monotone walk, and so induces a hazard. ■

Given this, the set of functions which contain only monotone walks is of interest.

Definition 5.3.3 *A logic function is said to be* **statically hazard-free** *iff each valid sequence on the cube is monotone.*

An interesting question that arises is the determination of necessary and sufficient conditions for a function to be statically hazard-free. Intuitively, we expect that a necessary condition is that the ways in which it can change value are highly restricted. We examine the conditions for which walks are monotone.

Lemma 5.3.1 *Let u be any arbitrary point on the n-cube. Then for every point v there is a valid sequence, $\{v, ..., u, ..., \overline{v}\}$.*

Proof: Induction on n, the dimensionality of the cube. For $n = 0$, trivial. Now suppose true for $n < N$, and consider the problem on the N-dimensional cube. If $u \neq v$, $u \neq \overline{v}$ [2], then there is an $N - 1$ dimensional face of the $N-$ cube, distance 1 from v, containing both u and $\overline{v}$. Let w be the unique point on this face distance 1 from v. Now, every valid sequence $\{w, ..., \overline{v}\}$ is a suffix of a valid sequence $\{v, w, ..., \overline{v}\}$, and is confined to the $N - 1$ dimension face containing w, u and $\overline{v}$. Now, by induction there is at least one valid sequence $\{w, .., u, ..., \overline{v}\}$, and hence there is a valid sequence $\{v, w, .., u, ..., \overline{v}\}$. ∎

This lemma leads immediately to strong characterization of the statically hazard-free functions.

Theorem 5.3.2 *Let f be a non-trivial statically hazard-free function on the n-cube c. Then for each vertex v_i of c, $f(v_i) \neq f(\overline{v_i})$.*

Proof: Suppose $f(v_i) = f(\overline{v_i})$ for some v_i. Since f is non-trivial there exists some vertex u such that $f(u) \neq f(v_i)$. By lemma 5.3.1, there is some valid sequence $\{v_i, ..., u, ..., \overline{v_i}\}$, and since $f(v_i) = f(\overline{v_i})$, this is non-monotone. ∎

Corollary 5.3.3 *On the n-cube, $f = 0$ on precisely half of the 2^n vertices for all statically hazard-free f.*

Proof: Follows immediately, for there is exactly one $\overline{v}$ for each vertex v. ∎

Corollary 5.3.4 *If f is a statically hazard-free function of $n > 1$ variables, then each face of the n-cube of dimension $n - 1$ contains at least one vertex where $f = 0$, and at least one where $f = 1$.*

Proof: Each face of dimension $n - 1$ contains half the points on the n-cube. If one such face consists entirely of zeroes (ones), the opposing face must consist entirely of ones (zeroes). The face containing all ones corresponds to a literal x, and hence the function f is isomorphic to the one-variable function x. ∎

---

[2] The problem is trivial if $u = v$ or $u = \overline{v}$

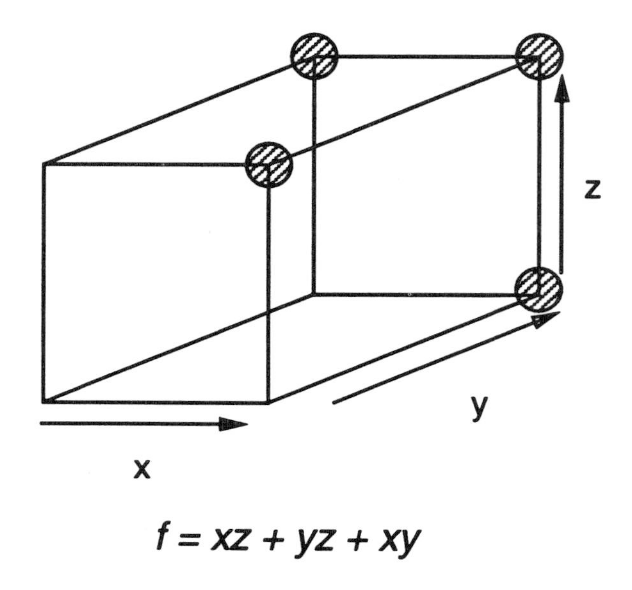

Figure 5.5: Only Function on the 3-cube Satisfying Corollaries 5.3.3-5.3.4

The set of functions satisfying the above theorem and corollaries is very small. Indeed, on the three-cube there are precisely two, one the complement of the other. One of these is shown in figure 5.5. Note that the walk $\{x\overline{yz}, x\overline{y}z, \overline{xy}z\}$ is non-monotone. Hence we conclude that no (non-trivial) function on the 3-cube is statically hazard-free, for this is the only possible statically hazard-free function on the 3-cube. In fact, this statement applied to the $n-$cube is true for every $n > 1$. We show this now.

Consider again the function in figure 5.5. Note, first, that every path beginning at $\vec{0}$ is monotone, and, second, that the function is *nondecreasing* in each variable; there is no place on the cube in which changing a variable from 0 to 1 changes the value of the function from 1 to 0. We now prove that every hazard-free function must be either nonincreasing or nondecreasing in each variable. The proof follows, but we state the intuition clearly here. If one considers the function as imposing a topography on the cube, then if we begin a walk at some vertex v, then v must be in a single basin of this topography (if $f(v) = 0$), or on a single plateau (if $f(v) = 1$).

Definition 5.3.4 *A function f is* **nondecreasing (nonincreasing)** *in* *a variable* x_j *iff changing* x_j *from 0 to 1 (1 to 0) does not change* f *from 1 to 0 (0 to 1). If* f *is either nonincreasing or nondecreasing in* x_j*, then* f *is said to be* **unate** *in* x_j*. Otherwise* f *is said to be* **binate** *in* x_j*. If* f *is nondecreasing (nonincreasing) in every variable, then* f *is said to be nondecreasing (nonincreasing). If* f *is either nondecreasing or nonincreasing in every variable, then* f *is said to be unate. Otherwise* f *is said to be binate.*

Theorem 5.3.5 *Let* f *be any function, and* $v = (x_1, ..., x_n)$ *be any vertex of the n-cube such that every valid sequence beginning at* v *is monotone. Let* $f(v) = 0$*. Then if* $x_j = 0$ *at* v*,* f *is nondecreasing in* x_j*, and if* $x_j = 1$ *at* v*, then* f *is nonincreasing in* x_j*. Similarly, if* $f(v) = 1$*, then if* $x_j = 0$ *at* v*,* f *is nonincreasing in* x_j*, and if* $x_j = 1$ *at* v*, then* f *is nondecreasing in* x_j*.*

Proof: We prove for the case $f(v) = 0$; the case $f(v) = 1$ follows by symmetry. Let $x_j = 0$ at v. If f is not nondecreasing in x_j, then there is some vertex $u = (.., x_j = 0, ...)$ such that $f(u) = 1$ and its neighbour vertex $u' = (.., x_j = 1, ...)$ such that $f(u') = 0$. There is a valid walk $\{v, ..., u, u', ..., \overline{v}\}$, and, since $f(v) = f(u') = 0, f(u) = 1$, $\{v, ..., u, u', ..., \overline{v}\}$ is non-monotone. Similarly, if $x_j = 1$ at v, then if f is not nonincreasing in x_j, then there is some vertex $u = (.., x_j = 1, ...)$ such that $f(u) = 1$ and its neighbour vertex $u' = (.., x_j = 0, ...)$ such that $f(u') = 0$. There is a valid walk $\{v, ..., u, u', ..., \overline{v}\}$, and, since $f(v) = f(u') = 0, f(u) = 1$, $\{v, ..., u, u', ..., \overline{v}\}$ is non-monotone. ∎

Corollary 5.3.6 *If* f *is statically hazard-free, then* f *is nondecreasing or* f *nonincreasing in every variable.*

Proof: Immediate, since if $f(\vec{0}) = 0$, then by the theorem f is nondecreasing in every variable, or $f(\vec{0}) = 1$, in which case f is nonincreasing in every variable. ∎

It is almost immediate now that there are no non-trivial statically hazard-free functions of greater than one variable. Consider vertices v and $\vec{0}$, $v \neq \vec{0}$, $f(v) = f(\vec{0})$, where every walk from either v or $\vec{0}$ is monotone. Note that f must remain nondecreasing (nonincreasing) on the cube obtained by rotating v into $\vec{0}$, which in turn implies that f is independent of at least some dimensions of the $n-$cube. We formalize this argument below, showing that v must be indistinguishable from $\vec{0}$.

We begin by demonstrating a sufficient condition for f to be independent of x.

Lemma 5.3.2 *If f is both nonincreasing and nondecreasing in some variable x, then f is independent of x*

Proof: Changing the value of x from 0 to 1 cannot change the value of the function either from 1 to 0 or from 0 to 1. Hence changing the value of x cannot change the value of the function, and done. ∎

Lemma 5.3.3 *Let v be any vertex, $f(v) = f(\vec{0})$, s.t. every valid sequence $\{\vec{0}, ..., \vec{1}\}$ and every valid sequence $\{v, ..., \overline{v}\}$ is monotone. Then f is independent of every variable x_j in which v differs from $\vec{0}$.*

Proof: WLOG, $f(\vec{0}) = f(v) = 0$. By the previous theorem, since $x_j = 0$ at $\vec{0}$ f must be nondecreasing in x_j. However, since $x_j = 1$ at v, f must be nonincreasing in x_j. Hence f is independent of x_j. ∎

Lemma 5.3.4 *Let v be any vertex, $f(v) = f(\vec{1})$, s.t. every valid sequence $\{\vec{1}, ..., \vec{0}\}$ and every valid sequence $\{v, ..., \overline{v}\}$ is monotone. Then f is independent of every variable x_j in which v differs from $\vec{1}$.*

Proof: Follows exactly the proof of lemma 5.3.3 ∎

Theorem 5.3.7 *Let f be a non-trivial function of $n > 1$ variables. Then f is not statically hazard-free.*

Proof: Induction on n. If $n = 2$, follows by case analysis on the functions $xy, x \oplus y, x + y$ (the other 7 true two-variable functions are isomorphic to one of these three for this purpose). Suppose theorem holds for $n < N$. If $n = N$, consider the set of N-variable functions which are not isomorphic to $N-1$ variable functions. If f is a statically-hazard-free function, then for every vertex v every valid sequence $\{v, ..., \overline{v}\}$ is monotone. In particular, every sequence $\{(1, 0, ..., 0), ..., (0, 1, .., 1)\}$ is monotone. By lemma 5.3.3, therefore, f is independent of x_0, contradicting the assumption that f was not isomorphic to a function of $N - 1$ variables. ∎

Theorem 5.3.8 *Let f be a non-trivial function of all of its variables. Then there is at most one vertex v such that every walk commencing at v is monotone and $f(v) = 0$, and at most one vertex v' such that every walk commencing at v' is monotone, and $f(v') = 1$. Further, if such a vertex v exists, then v' exists and $v' = \overline{v}$.*

Proof: That at most one such v and one such v' exist is immediate from lemma 5.3.3, so all that must be shown is that if v exists, v' exists and $v' = \overline{v}$. For this, suppose v exists. By appropriate change of variables we can ensure that $v = \vec{0}$, and hence that the function is nondecreasing in all its variables. But immediately, then, we have that $f(\vec{1}) = 1$, and that every walk from $f(\vec{1})$ is monotone, and of course $\vec{1} = \overline{\vec{0}}$, and so done. ∎

The preceding theorems show that if a function is not to have a hazard under evaluation, then the function must be unate, and, further, (after some rotation of the cube) the evaluation must begin at the state $\vec{0}$ or $\vec{1}$. The practice of setting each variable to a known state before evaluation is known as *precharging*.

Thus precharging and unateness are necessary for hazard avoidance. The question is, are they sufficient? The following theorem provides this final piece to the puzzle.

Theorem 5.3.9 *If f is nondecreasing, then every sequence $\{\vec{0}, ..., \vec{1}\}$ is monotone.*

Proof: f is nondecreasing. Let $P = \{\vec{0}, v_1, ..., v_i, ..., \vec{1}\}$ be any valid sequence. We must show it is monotone.

Let v_i be the first node such that $f(v_i) = 1$. Then we must show $f(v_j) = 1 \, \forall j > i$. We proceed by induction. For v_{i+1}, observe that the only difference between v_i and v_{i+1} is that some variable, say x_k, was changed from 0 to 1. This cannot change f to 0, since f is nondecreasing. Similarly, if $f(v_j) = 1 \, \forall i < j < N$, the only difference between v_N and v_{N-1} is that variable x_r for some r was changed from 0 to 1 and hence, $f(v_N) = f(v_{N-1}) = 1$, and done. ∎

In sum, in this section we have shown that the family of hazard-free circuits, under our initial assumptions, is identical to the family of precharged unate circuits. In the next two sections, we will explore two of our base assumptions, and show that relaxing either or both assumptions has no effect on the results, and in the second case that the assumption cannot be easily relaxed in a VLSI environment.

5.4 The SDC Set and Restricted Cubes

The preceding arguments rested on assumption (5): any starting or ending vertex was permitted. In practice, this is not the case. If the input variables to a node are themselves functions of the primary input variables, then various combinations of the input variables (typically called *faces* or *cubes*) may not be possible *static* values of the input vertices, and if so, are not appropriate terminal vertices of a walk. The collection of such impossible cubes is generally known as the *Satisfiability Don't-Care (SDC) Set*[3]. Now, even for non-precharged logic one can view the set of *allowable* starting and terminal vertices to reside in $\overline{SDC}$. However, it would be an error to consider that all walks across a cube must avoid the SDC set, for the vertices within these cubes are only forbidden as *static* values; there is no reason to believe that the vertices within these cubes may not occur as transient values as the inputs change.

Nevertheless, certain cubes may be forbidden. Given any start vertex v(any vertex from which all walks are monotone), any cube of the SDC set containing the terminal vertex $\overline{v}$ will not be entered on any valid walk from v, since such a walk will not exit the cube of the SDC set and hence will not terminate at a statically valid vertex. We call these cubes the *restricted cubes* of vertex v, denoted $R(v)$.

Definition 5.4.1 *A cube c is a* **restricted vertex** *of v iff $c \subseteq SDC$ and $\overline{v} \in c$. The set of all restricted vertices of v is denoted $R(v)$.*

The essential point about the vertices contained within the Satisfiability Don't Care set is that the value of the function may be chosen arbitrarily on these points, since these values will never be realized statically. The implemented function, of course, is completely specified; there is a real, concrete value attached to each point on the $n-$cube. The concrete value is a matter of concern for those vertices inside the SDC set but outside $R(v)$, since these will be visited in transit. Within $R(v)$, however, we can choose values arbitrarily, since these vertices will never be visited. The strategy we use to prove results in this section is to demonstrate that a single "good" assignment of function values exists for vertices in $R(v)$, and that by making that assignment we do not impose any restrictions on the hazard freedom of the realized function. We then show that the results of the preceding sections hold for such functions; this in turn shows that redundancy information in the form of the SDC set does not yield any significant loosening of the precharged, unate requirements of the previous sections. The "good" assignment simply forces values in the unreachable parts of the cube ($R(v)$) to be the value opposite that of the function at v.

INTEGRATING FUNCTIONAL AND TEMPORAL DOMAINS IN LOGIC DESIGN

THE FALSE PATH PROBLEM AND ITS IMPLICATIONS

THE KLUWER INTERNATIONAL SERIES IN ENGINEERING AND COMPUTER SCIENCE

VLSI, COMPUTER ARCHITECTURE AND DIGITAL SIGNAL PROCESSING

Consulting Editor
Jonathan Allen

Latest Titles

VLSI Design for Manufacturing: Yield Enhancement, S.W..Director, W. Maly, A.J. Strojwas, ISBN: 0-7923-9053-7
Testing and Reliable Design of CMOS Circuits, N.K. Jha, S. Kundu, ISBN: 0-7923-9056-3
Hierarchical Modeling for VLSI Circuit Testing, D. Bhattacharya, J.P. Hayes, ISBN: 0-7923-9058-X
Steady-State Methods for Simulating Analog and Microwave Circuits, K. Kundert, A. Sangiovanni-Vincentelli, J. White, ISBN: 0-7923-9069-5
Introduction to Analog VLSI Design Automation, M. Ismail, J. Franca, ISBN: 0-7923-9102-0
Gallium Arsentide Digital Circuits, O. Wing, ISBN: 0-7923-9081-4
Principles of VLSI System Planning, A.M. Dewey , S.W. Director ISBN: 0-7923-9102-0
Mixed-Mode Simulation, R. Saleh, A.R. Newton, ISBN: 0-7923-9107-1
Automatic Programming Applied to VLSI CAD Software: A Case Study, D. Setliff, R.A. Rutenbar, ISBN: 0-7923-9112-8
Models for Large Integrated Circuits, P. Dewilde, Z.Q. Ning ISBN: 0-7923-9115-2
Hardware Design and Simulation in VAL/VHDL, L.M. Augustin, D.C..Luckham, B.A.Gennart, Y.Huh, A.G.Stanculescu ISBN: 0-7923-9087-3
Subband Image Coding, J. Woods, editor, ISBN: 0-7923-9093-8
*Low-Noise Wide-Band Amplifiers in Bipolar and CMOS Technologies,*Z.Y.Chang, W.M.C.Sansen, ISBN: 0-7923-9096-2
Iterative Identification and Restoration of Images, R. L.Lagendijk, J. Biemond ISBN: 0-7923-9097-0
VLSI Design of Neural Networks, U. Ramacher, U. Ruckert ISBN: 0-7923-9127-6
Synchronization Design for Digital Systems, T. H. Meng ISBN: 0-7923-9128-4
Hardware Annealing in Analog VLSI Neurocomputing, B. W. Lee, B. J. Sheu ISBN: 0-7923-9132-2
Neural Networks and Speech Processing, D. P. Morgan, C.L. Scofield ISBN: 0-7923-9144-6
Silicon-on-Insulator Technology: Materials to VLSI, J.P. Colinge ISBN: 0-7923-9150-0
Microwave Semiconductor Devices, S. Yngvesson ISBN: 0-7923-9156-X
A Survey of High-Level Synthesis Systems, R.A. Walker, R. Camposano ISBN: 0-7923-9158-6
Symbolic Analysis for Automated Design of Analog Integrated Circuits , G. Gielen, W. Sansen, ISBN: 0-7923-9161-6
High-Level VLSI Synthesis, R. Camposano, W. Wolf, ISBN: 0-7923-9159-4

INTEGRATING FUNCTIONAL AND TEMPORAL DOMAINS IN LOGIC DESIGN

THE FALSE PATH PROBLEM AND ITS IMPLICATIONS

by

Patrick C. McGeer
University of British Columbia

and

Robert K. Brayton
University of California

Kluwer Academic Publisher
Boston/Dordrecht/London

Distributors for North America:
Kluwer Academic Publishers
101 Philip Drive
Assinippi Park
Norwell, Massachusetts 02061 USA

Distributors for all other countries:
Kluwer Academic Publishers Group
Distribution Centre
Post Office Box 322
3300 AH Dordrecht, THE NETHERLANDS

Library of Congress Cataloging-in-Publication Data
McGeer, Patrick C.
Integrating functional and temporal domains in logic design : the false path problem and its implications / by Patrick C. McGeer and Robert K. Brayton.
p. cm. -- (The Kluwer international series in engineering and computer science ; 139, VLSI, computer architecture, and digital signal processing)
Includes bibliographical references and index.
ISBN 0-7923-9163-2
1. Logic design--Data processing. 2. Integrated circuits--Very large scale integration--Design construction--Data processing.
3. Computer-aided design. I. Brayton, Robert King. II. Title.
III. Series: Kluwer international series in engineering and computer science ; SECS 139. IV. Series: Kluwer international series in engineering and computer science. VLSI, computer architecture, and digital signal processing.
TK7868.L6M4 1991
621.39'5--dc20 91-14516
CIP

Printed on acid-free paper.

Printed in the United States of America

For Karen and for Ruth

Contents

List of Figures

Now, let V be the unique maximum subset of $S(f_i, P)$ such that $c \in \psi^{g,\tau_i - 1}$ for each $g \in V$. By lemma A.3.1 the fact that $c \in \psi^{g,\tau_i-1}$ can be determined in polynomial time for each g, and hence the determination of V is polynomial by the boundedness of $S(f_i, P)$. It is easy to demonstrate that c satisfies

$$\sum_{U \subseteq S(f_i,P)} (\mathcal{S}_U \frac{\partial f_i}{\partial f_{i-1}}) \prod_{g \in U} \psi^{g,\tau_i-1}$$

iff c satisfies

$$(\mathcal{S}_V \frac{\partial f_i}{\partial f_{i-1}}) \prod_{g \in V} \psi^{g,\tau_i-1}$$

and hence iff c satisfies

$$\mathcal{S}_V \frac{\partial f_i}{\partial f_{i-1}}$$

The calculation of $\mathcal{S}_V \frac{\partial f_i}{\partial f_{i-1}}$ is easily made in polynomial time, and, further, the determination that c satisfies $\mathcal{S}_V \frac{\partial f_i}{\partial f_{i-1}}$ is easily made in polynomial time by direct simulation. ■

This theorem, together with the theorems which demonstrate that LVP is $\mathcal{NP}$-hard, demonstrates that LVP is $\mathcal{NP}$-complete.

Appendix B

A Family of Operators

The Boolean difference and the smoothing operator, explored earlier, can be thought of as two members of a family of operators involving the cofactors. Each of these operators reduces the dimension of the space by one variable, but the semantics of the operators vary. The character of this family can be divined by examining the formulae for the boolean difference:

$$\frac{\partial f}{\partial x} = f_x \oplus f_{\overline{x}}$$

and of the smoothing operator:

$$\mathcal{S}_x f = f_x + f_{\overline{x}}$$

The hint here is that for each of the 16 two-variable boolean functions, there should be a separate cofactor operator. The purpose of this appendix is to enumerate these operators, describe their function and their interrelationship. This taxonomy is not particularly useful, but it does serve to beautify science.

The sixteen dyadic boolean functions are as described in table B.1. these functions correspond to operators as described in table B.2, letting f_x stand for x and $f_{\overline{x}}$ stand for y Now, it is relatively clear through an examination of these operators that these operators are related in a fairly rich way. In particular, consider the *duality* relation: an operator O is the dual of an operator O' iff $O_x f = {O'}_x \overline{f}$. Clearly duality is a symmetric relationship. We can write the duality table in table B.3, omitting trivial operators:

Note that only eight of the 16 operators are mentioned in the duality table. This follows from the observation that the operators 0, 1, f_x, $\overline{f_x}$,

x	y	0	1	$\overline{x}$	$\overline{y}$	xy	$x+y$	$\overline{xy}$	$\overline{x+y}$
0	0	0	1	1	1	0	0	1	1
0	1	0	1	1	0	0	1	1	0
1	0	0	1	0	1	0	1	1	0
1	1	0	1	0	0	1	1	0	0

$x \geq y$	$x > y$	$x \leq y$	$x < y$	$x \overline{\oplus} y$	$x \oplus y$
1	0	1	0	1	0
0	0	1	1	0	1
1	1	0	0	0	1
1	0	1	0	1	0

Table B.1: Dyadic Boolean Functions

Function	Operator	Semantic
x	f_x	f evaluated at $x = 1$
y	$f_{\overline{x}}$	f evaluated at $x = 0$
$\overline{x}$	$\overline{f_x}$	$\overline{f}$ evaluated at $x = 1$
$\overline{y}$	$\overline{f_{\overline{x}}}$	$\overline{f}$ evaluated at $x = 0$
0	0	0
1	1	1
xy	$\mathcal{C}_x f = f_x f_{\overline{x}}$	$f = 1$ for each value of x
$x + y$	$\mathcal{S}_x f = f_x + f_{\overline{x}}$	$f = 1$ for some value of x
$\overline{xy}$	$\overline{\mathcal{S}_x f}$	$f = 1$ for no value of x
$\overline{x+y}$	$\overline{\mathcal{C}_x f}$	$f = 0$ for some value of x
$x \geq y$	$\mathcal{I}_x f$	f is monotone increasing in x
$x > y$	$\mathcal{IP}_x f$	f is strictly monotone increasing in x
$x < y$	$\mathcal{DP}_x f$	f is strictly monotone decreasing in x
$x \leq y$	$\mathcal{D}_x f$	f is monotone decreasing in x
$x \oplus y$	$\frac{\partial f}{\partial x}$	f is determined by x
$x \overline{\oplus} y$	$\overline{\frac{\partial f}{\partial x}}$	f is independent of x

Table B.2: Dyadic Boolean Functions and their Corresponding Operators

Operator	Dual
$\mathcal{C}_x f$	$\mathcal{S}_x f$
$\mathcal{I}_x f$	$\mathcal{D}_x f$
$\mathcal{IP}_x f$	$\mathcal{DP}_x f$
$\frac{\partial f}{\partial x}$	$\frac{\partial f}{\partial x}$
$\frac{\partial f}{\partial x}$	$\frac{\partial f}{\partial x}$

Table B.3: Duality Table of Operators

$f_{\overline{x}}$, $\overline{f_{\overline{x}}}$ are trivial, and that the duality properties of the operators $\overline{\mathcal{S}_x f}$ and $\overline{\mathcal{C}_x f}$ are adequately captured elsewhere in the table.

Appendix C

Fast Procedures for Computing Dataflow Sets

C.1 Introduction

In computing whether paths are true by some sensitization criterion, we assert values of nodes in a multi-level network and then determine whether some input vector can **justify** the assertions that we make. If such a vector exists, we say that the multi-level function implicitly expressed by these assertions is **satisfiable**. In previous appendices, we have seen that in general this is a hard problem.

Now, it is clear that any function which requires both that y and $\bar{y}$ be true for some variable y cannot be satisfiable; such a function is said to have an **explicit incompatibility**. Hence one approach to the satisfiability problem is simply to determine whether a function has an explicit incompatibility, and delare it unsatisfiable only if it does. Note that this is an inexact procedure: a function may not have an explicit incompatibility but still be unsatisfiable. Hence this is a *biased* SAT test: all satisfiable functions are reported as satisfiable, but some unsatisfiable functions may be reported as satisfiable. However, referring to the approximation spectrum in 4.2, it is clear that this is a *positively-biased SAT test*, which, as we reported there, is a safe approximation to SAT: using this in a false-path detection algorithm will not result in a true path being reported as false. This appendix details a fast ($O(n^4)$) procedure for computing such a safe approximation. It computes the implications of any assertion.

Dataflow computations for optimizing Boolean logic networks are well known [16, 8, 85, 37]. Such analysis computes inferences of the form $y_i = b_i \Rightarrow y_j = b_j$ for nodes y_i, y_j and b_i, b_j in $\{0, 1\}$. In the literature, the set $\mathcal{F}_{ij}(x)$ is used to represent the set of nodes y_k s.t. y_k is set to j when x is set to i.

Now, a polynomial algorithm to compute these sets exactly for general networks obviously implies a polynomial solution to the co-$\mathcal{NP}$-complete problem of tautology (setting any primary input to either 1 or 0 sets the output x to 1 iff the network is tautologous, and hence $x \in \mathcal{F}_{11}(y_k) \cap \mathcal{F}_{01}(y_k) \forall y_k$, where y_k is a primary input iff the network is tautologous).

For this reason, the sets $\mathcal{F}_{ij}$ are not computed. Berman, Trevillyan and Joyner [8] have defined, on networks of NOR gates, the interesting subsets $C_{ij}(x) \subseteq \mathcal{F}_{ij}(x)$. These are defined by the rules:

$$y \in C_{10}(x) \quad \text{if} \quad \exists t \in C_{11}(x) \text{ and } y \text{ is a fanout of } t \tag{C.1}$$

$$y \in C_{10}(x) \quad \text{if} \quad \exists t \in C_{11}(x) \text{ and } y \text{ is a fanin of } t \tag{C.2}$$

$$y \in C_{10}(x) \quad \text{if} \quad \begin{array}{l} \exists t \in C_{10}(x) \text{ and } y \text{ is the only fanin} \\ \text{of } t \text{ not in } C_{10}(x) \end{array} \tag{C.3}$$

$$y \in C_{10}(x) \quad \text{if} \quad \exists t \in C_{11}(x) \text{ and } y \in C_{10}(t) \tag{C.4}$$

$$y \in C_{10}(x) \quad \text{if} \quad \exists t \in C_{10}(x) \text{ and } y \in C_{00}(t) \tag{C.5}$$

$$y \in C_{10}(x) \quad \text{if} \quad x \in C_{10}(y) \tag{C.6}$$

The subsets are computed by finding the least fixed point of the recurrence relations given by these rules and the analogous rules for C_{11}, C_{00}, and C_{01}. Initially, $C_{11}(x) = x$, $C_{00}(x) = \overline{x}$, and $C_{01}(x) = C_{10}(x) = \emptyset$ for all x.

The four rules are generally self-explanatory. The first rule (the *fanout* rule) captures the fact that in a network of nor gates, setting $t = 1$ sets all fanouts of t to 0. The second *fanin* rule is derived from the observation that if t is set to 1, then all of its fanins are set to 0. The third (another fanin) rule captures the fact that if $t = 0$, at least one of its inputs must be 1. Rules 4 and 5 transitively close the sets. Rule 6 captures the well-known rule of deduction $x \Rightarrow \overline{y}$ is equivalent to $y \Rightarrow \overline{x}$

While this approach has demonstrated some power, nevertheless some improvements are desirable. First, it is tedious to develop new rules for each sort of gate, and for combinations of gates. Second, one wishes an explicit algorithm for the computation of these sets, with some hope that it is efficient.

This appendix develops a fast procedure to compute these sets, and a generalization to all boolean networks.

C.2 Terminology

Recall from chapter 1 that a product of literals is called a *cube.* A cube may also be viewed as a *set* of literals; the cube xyz is equivalent to the set $\{x, y, z\}$. This equivalence of sets and cubes permits us to regard the sets $C_{ij}(x)$ as cubes $C_{ij}(x)$. This permits a new, generalized approach to the derivation of C-sets.

In the derivation, we will be using the cofactor notation a great deal; we remind the reader here that f_c refers to the function f evaluated on the space defined by the cube c.

C.3 The New Approach

The traditional division of the dataflow implications into four sets is an artifact of the traditional nor- or nand-gate formulation. In fact, for each x, the sets of interest are the sets of literal values implied by choosing either $x = 1$ or $x = 0$. By using both phases, one can take the union of $C_{10}(x)$ and $C_{11}(x)$ as the set C_x (the set of values implied by $x = 1$), and the union of $C_{00}(x)$ and $C_{01}(x)$ as the set $C_{\overline{x}}$ (the set of values implied by $x = 0$).

We define C_x ($C_{\overline{x}}$) as the fixed points of the set sequence C_x^i ($C_{\overline{x}}^i$), where $C_x^0 = \{x\}$, $C_{\overline{x}}^0 = \{\overline{x}\}$, and C_x^{n+1} is obtained from C_x^n by the relations:

$$y \in C_x^n \quad \text{if} \quad y_{C_x^{n-1}} = 1 \tag{C.7}$$

$$y \in C_x^n \quad \text{if} \quad \exists t \in C_x^{n-1} \text{ and } t_{C_x^{n-1}} = yf \text{ some } f \tag{C.8}$$

$$y \in C_x^n \quad \text{if} \quad \exists \overline{t} \in C_x^{n-1} \text{ and } t_{C_x^{n-1}} = \overline{y} + f \text{ some } f \tag{C.9}$$

$$y \in C_x^n \quad \text{if} \quad \exists t \in C_x^{n-1} \text{ and } y \in C_t^{n-1} \tag{C.10}$$

$$y \in C_x^n \quad \text{if} \quad \overline{x} \in C_{\overline{y}}^n \tag{C.11}$$

And, of course, the symmetries obtained by substituting $\overline{y}$, and/or $\overline{t}$ for t, and/or $\overline{x}$ for x. Relation (C.7) is the usual fanout rule; relations

(C.8) and (C.9) capture the fanin rule; and relation (C.10) is transitiv closure. Relation (C.11) is the contrapositive. Note that these rule simplify to the well-known nor-gate rules for networks consisting only o nor gates.

Lemma C.3.1 *Let $y(\overline{y}) \in C_x$, C_x is obtained as a fixed point of relation (C.7)-(C.11). Then y is set to 1 (0) whenever x is set to 1, and th similar observation holds for $C_{\overline{x}}$.*

Proof: WLOG, we consider y only in the positive phase, and x in th positive phase; the other three cases follow by symmetry. We prove b induction on n, the level at which y is added to the set C_x (that is $y \in C_x^n - C_x^{n-1}$. The base case is trivial (x is clearly set to 1 when x i set to 1), so suppose the statement holds for all $z \in C_w^{n-1}$, $\forall w$. Now, is added at n, and must be added by one of (C.7)-(C.11). If by (C.7) then $y_{C_x^n} = 1$, and since the settings are correct in C_x^n by the inductiv assumption, y is certainly set to 1 when x is set to 1. If by (C.8), the $\exists t \in C_x^{n-1}$ and $t_{C_x^{n-1}} = yf$. By the inductive assumption t is set to when x is set to 1, and $t_{C_x^{n-1}} = yf$, hence for $t = 1$ we must have $y = 1$ whence we must set y to 1. If by (C.9), then $\exists \overline{t} \in C_x^{n-1}$ s.t. $t_{C_x^{n-1}} = \overline{y} + f$ For $t = 0$, as required, we must have $\overline{y} = 0$, whence y must be set to 1 If by (C.10), then we have $w = 1$ from $x = 1$ and $y = 1$ from $w = 1$ whence $x = 1$ implies $y = 1$. If by (C.11), we have that $y = 0 \Rightarrow x = 0$ Hence if $x = 1$ we must have $y = 1$, for if $y = 0$ then $x = 0$, and so done ■

We now turn to the computation of the sets. The details of thi computation are important for the efficiency of the algorithm.

C.4 Computations

The preliminary observation that we make before we begin the algo rithms is the duality between cubes and sets. Using this duality, we ca store the sets C_x^n and $C_{\overline{x}}^n$ as cubes, and use the logic operations of co factoring for node evaluation and boolean AND to find the union of tw sets.

In this and subsequent code, it is important to understand precisel the difference between a *variable* and a *literal*. Strictly speaking, a litera is the instance of a variable in either of its phase; it may be thought of a a pair (variable, phase), where phase is in {0,1}. By abuse of notation, literal is generally represented by the appropriate variable in its positive phase; thus is brevity the enemy of precision.

We keep to this convention here. Except where explicitly noted, all arguments to the functions developed below are literals, and hence may be in either phase, though they will always appear, by convention, in positive phase. Further, the sets C^n_x, which will be represented by the variables $\mathcal{C}_x$, are indexed by literals and not variables.

C.4.1 Basic Algorithms

The two fundamental procedures are the fanout and fanin evaluation procedures. The former attempts to discover nodes that are set to a constant under the cofactoring operation; the latter attempts to discover nodes which have non-trivial cube factors under the cofactoring operation.

The fanout evaluation procedure returns 0, 1, or 2, according as to whether y is set to 0, 1, or neither by the cofactoring process. y in this code is a node, not a literal.

```
evaluate_node(y, x)
{
    df_cube = C_x;
    eval_node = cofactor(y, df_cube);
    if(eval_node == 1) return 1;
    else if(eval_node == 0) return 0;
    else return 2;
}
```

The fanin procedure returns the common cube dividing the cofactored cube. The process is relatively straightforward: the literal y is known to be set when the literal x is set. For the moment, assume that the phase of y is positive. If y is set to a product of a `cube` and some function by the cofactoring process (in other words, when the cubes of the cofactored node have a non-trivial intersection), then the literals of `cube` must be set appropriately to set `cube` to 1.

Now consider the case where y is in its negative phase. Hence we must have $\overline{\mathrm{y}} = 1$ under $\mathcal{C}_x$, and using the fact that $\overline{(f_c)} = \overline{f}_c$, we can run the described algorithm on $\overline{\mathrm{y}}$

```
evaluate_fanin(y, x)
{
    if(y is in positive phase) eval_node = y_{C_x};
    else eval_node = \bar{y}_{C_x};
    fanin_cube = get_common_cube(eval_node);
    return common_cube;
}
```

C.4.2 Transitivity

The results of the preceding section suffice to capture the fanin an fanout rules, respectively. There remains the matter of transitive closur Immediate transitivity can be guaranteed by rewriting the fanout an fanin rules as follows:

$$C_x^{n+1} \supseteq C_y^n \quad \text{if} \quad y_{C_x^n} = 1 \tag{C.12}$$

$$C_x^{n+1} \supseteq C_y^n \quad \text{if} \quad \exists t \in C_x^n \text{ and } t_{C_x^n} = yf \text{ some } f \tag{C.13}$$

(with the usual symmetries). The contrapositive rule is rewritten:

$$C_x^{n+1} \supseteq C_y^n \quad \text{if} \quad \overline{x} \in C_{\overline{y}}^n \tag{C.14}$$

Put bluntly, the entire set of literals C_y^n is included in C_x^{n+1}, rathe than simply y. The reader can easily verify that this is correct as a immediate consequence of the previously-given transitive closure rule This is accomplished by the following code:

```
merge_df_cubes(x, y)
{
    if(C_x ⊇ C_y) return 0;
    else {
        C_x = C_x ∪ C_y;
        return 1;
    }
}
```

Note that `merge_df_cubes` returns 1 only if $C_x^{n+1} \neq C_x^n$.

Let us quickly consider the matter of the contrapositive. This ca be handled most naturally by merging $C_{\overline{x}}$ into $C_{\overline{y}}$ whenever C_y is merge into C_x, as implied by (C.14).

There remains the matter of further transitivity. Certainly C_x^{n+1} is :ansitively closed by the operations given above. However, if x appears ı the positive phase in C_z^n for some z, then the transitivity closure ıles require that $C_z^{n+2} \supseteq C_x^{n+1}$. Now, if $C_x^n = C_x^{n+1}$, this is assured ıductively. The difficulty arises if C_x^{n+1} has changed. In this case, we ıust propagate its change to all of the dataflow cubes where x appears ı the positive phase. This is done by maintaining sets of cubes D_x and $D_{\overline{x}}$ for each node x, where D_x is the set of cubes C_y containing x, and $D_{\overline{x}}$ the cubes C_z containing $\overline{x}$.

```
ropagate_dataflow(x)

    foreach y ∈ D_x {
        merge_df_cubes(y, x);
        merge_df_cubes(x̄, ȳ);
    }
```

The maintenance of the sets D_z requires a change to `merge_df_cubes`; ›r if z is in C_y^n and is not in C_x^n, then C_x^{n+1} must be added to D_z.

```
erge_df_cubes(x, y)

    if(C_x ⊇ C_y) return 0;
    else {
        C_x = C_x ∪ C_y;
        foreach z ∈ C_y
            D_z = D_z ∪ {C_x}
        return 1;
    }
```

:.4.3 When $C_x = 0$

'here remains the matter of the interpretation of zeroing one of the C_x^i. This occurs iff, for some variable y, $y \in C_x^i$ and $\overline{y} \in C_x^i$; hence $= 1 \Rightarrow y = 1$ and $y = 0$. This is impossible, hence if $C_x^i = 0$, then cannot be set to 1. Since x cannot be set to 1, then all of the other ariable settings which imply $x = 1$ are also impossible, and hence their corresponding cubes must be set to 0. These are the cubes contained in the transitive closure of D_x.

```
propagate_zero(x)
{
    stack = tfo_collect(x);
    while((z = pop_stack(stack)) != nil)
        C_z = 0;
}
```

The transitive-fanout collection procedure is a standard graph traversal through the edges implied by the sets D_y. `propagate_zero` is called from `propagate_dataflow` if the propagated cube is ever found to be 0.

C.4.4 Evaluation Algorithms

There remains the question of evaluation. Literals may be added to a C set either through propagation or evaluation. Propagation has been adequately covered above. We turn to evaluation. The core of the evaluation strategy is in the following lemma and definition.

Definition C.4.1 *If some variable y is set to 1 (0) by either the routine* `evaluate_node` *or the routine* `evaluate_fanin`*, under the implications of some dataflow cube C_x^n, we say that y is* **implied** *by C_x^n.*

Now, clearly, at each iteration of the evaluation algorithm, we only wish to examine the variables which *may* be implied by C_x^n. We isolate these *potential implicants* of C_x^n by the following lemma.

Lemma C.4.1 *Let y be an implicant of C_x^n, y is* not *an implicant of C_x^{n-1}. Then y is a fanin or a fanout of some literal in $C_x^n - C_x^{n-1}$, or a fanin of some literal in C_x^n which has a fanin in $C_x^n - C_x^{n-1}$.*

Proof: y is an implicant of C_x^n only if it is set to some value under C_x^n by `evaluate_node` or by `evaluate_fanin`. In the former case, it must be a fanout of some node in C_x^n, since cofactoring by a cube only may set the values of fanouts of that cube. Further, if y is not a fanout of some literal in $C_x^n - C_x^{n-1}$ $y_{C_x^n} = y_{C_x^{n-1}}$, a contradiction since y is not an implicant of C_x^{n-1}. In the latter case, then there is some literal $z \in C_x^n$ s.t. $z_{C_x^n} = yf$, some f (or $\overline{z} \in C_x^n, z_{C_x^n} = \overline{y} + f$. Now, since y not an nplicant of C_x^{n-1}, either z not in C_x^{n-1} in which case $z \in C_x^n - C_x^{n-1}$, or $C_x^{n-1} \neq z_{C_x^n}$, in which case at least one fanin of z in $C_x^n - C_x^{n-1}$. ∎

With this in hand, we can proceed with the evaluation algorithm. he previous lemma suggests an event-driven approach. We maintain stack of records, each record containing a dataflow cube and a literal ewly added to the cube. At each iteration, one such record is popped ff the stack and the potential new implicants of the dataflow cube are xamined. These are, by lemma, the fanouts of the new literal, the fanins f the new literal, and the other fanins of the fanouts of the new literal. his is captured by the following code:

```
hile((Cx, y) = pop_stack(evaluation_stack)) {
   if Cx = 0 continue;
   foreach fanout w of y {
        if((w has a literal w1 ∈ Cx) { (w1 is either w or w̄)
           new_cube = evaluate_fanin(w1, x);
           foreach literal z in new_cube
               merge_df_cubes(x, z);
               merge_df_cubes(z̄,x̄);
        }
        phase = evaluate_node(w, x);
        if(phase != 2){
           w1 is the literal suggested by w and phase;
           merge_df_cubes(x, w1);
           merge_df_cubes(w1̄,x̄);
        }
   }
   new_cube = evaluate_fanin(y, x);
   foreach literal z in new_cube
        merge_df_cubes(x, z);
        merge_df_cubes(z̄,x̄);
   propagate_dataflow(x);
```

`merge_df_cubes` is modified to add new elements to this stack, as terals are added to the dataflow cubes. The stack is initially a set of ecords of the form (C_X, x) for each node x. The algorithm terminates hen the stack is empty.

C.5 Correctness

We now turn to a proof of correctness of the algorithms given above. The correctness of `evaluate_node`, `evaluate_fanin`, `merge_df_cubes`, `propagate_dataflow`, and `propagate_zero` is evident, or has been adequately treated above. We now establish the correctness of the package.

Lemma C.5.1 *Let $x \in \mathcal{C}_y$. Then either $\mathcal{C}_y \supseteq \mathcal{C}_x$ or $\mathcal{C}_x$ is on the stack.*

Proof: We construct a loop invariance argument. Clearly on the 0th iteration the statement holds. Suppose it holds through k iterations. Now suppose $x \in \mathcal{C}_y$ on the $k+1$st iteration. If the statement of the lemma does not hold through this iteration, then either $\mathcal{C}_x$ was popped off the stack, or $\mathcal{C}_x$ grew and $\mathcal{C}_y$ did not grow to contain it, or $\mathcal{C}_x$ was added to $\mathcal{C}_y$ and not all of $\mathcal{C}_x$ was added to $\mathcal{C}_y$. In the first case, at the end of the iteration `propagate_dataflow(x)` was called. Since $\mathcal{C}_y \in D_x$, when `propagate_dataflow(x)` $\mathcal{C}_y$ is updated to contain $\mathcal{C}_x$. In the second case, when $\mathcal{C}_x$ grows so that it no longer is contained $\mathcal{C}_y$, $\mathcal{C}_x$ is shoved on the stack. In the third case, there is some $\mathcal{C}_z$, $z \neq x$ such that $x \in \mathcal{C}_z$, $\mathcal{C}_x \not\subseteq \mathcal{C}_z$ was added to $\mathcal{C}_y$. But hence, on the kth iteration, we must have had $\mathcal{C}_x$ on the stack by the invariance assumption. ∎

Corollary C.5.1 *At the completion of the algorithm, if $x \in \mathcal{C}_y$, $\mathcal{C}_y \supseteq \mathcal{C}_x$.*

Theorem C.5.2 *Let C_x be the fixed points of relations (C.7)-(C.11). Then the final value of $\mathcal{C}_x = C_x$ $\forall x$*

Proof: $\mathcal{C}_x \supseteq C_x$. We construct an inductive argument on n, and show that $\mathcal{C}_x \supseteq C_x^n$ for every n. Since $C_x = C_x^n$ for some n, this gives the result. Clearly $\mathcal{C}_x \supseteq C_x^0 (= \{x\})$. Assume that $\mathcal{C}_x \supseteq C_x^n$ $\forall n \leq N$. Let $y \in C_x^{N+1}$. Now, either $y \in C_x^N$, and done, or y was added by one of relations (C.7)-(C.11). If by (C.7), then $y_{C_x^N} = 1$. Since $\mathcal{C}_x \supseteq C_x^N$ by assumption, we must have that $y_{\mathcal{C}_x} = 1$ for all iterations through the algorithm after the last element of *fanins*$(y) \cap C_x^N$ (call this z) was added to $\mathcal{C}_x$. However, once z was added to $\mathcal{C}_x$, then $(z, \mathcal{C}_x)$ was pushed on the evaluation stack by `merge_df_cubes`. When it was subsequently popped, all the fanouts of z were examined, including y. Since $y_{\mathcal{C}_x} = 1$ for that value of $\mathcal{C}_x$, $\mathcal{C}_y$ is merged with $\mathcal{C}_x$. Since $y \in \mathcal{C}_y$, we are done for this case. The cases of (C.8)-(C.9) are shown by similar arguments. For the case where y is added by (C.10), let t be the literal in the statement

of the relation (C.10). By assumption $t \in C_x$ and $y \in C_t$. Now, either t is added to C_x after y is added to C_t, in which case y is added to C_x when C_t is merged into C_x in the main loop of the algorithm, or y is added to C_t after t is added to C_x, in which case $C_x \in D_t$ when y is added to C_t, and so y is added to C_x when C_t is merged into C_x in `propagate_dataflow`. For the case where y was added by (C.11), then on some iteration $\overline{x}$ was added to $C_{\overline{y}}$ (by induction, since $\overline{x} \in C_{\overline{y}}^N \subseteq C_{\overline{y}}$), and $C_{\overline{y}}$ was added to $D_{\overline{x}}$, and $(\overline{x}, C_{\overline{y}})$ was pushed on the stack. When it is popped, `propagate_dataflow(`$\overline{y}$`)` will merge $C_{\overline{x}}$ into $C_{\overline{y}}$, and then C_y into C_x, adding y to C_x, and done.
$C_x \subseteq C_x$. We show this by a loop invariance argument. Clearly $C_x \subseteq C_x$ through 0 iterations of the loop. Suppose that $C_x \subseteq C_x$ through k iterations of the loop; we must show that $C_x \subseteq C_x$ through the $k+1$st iteration. Suppose that y, C_x is the pair popped off the evaluation stack at the $k+1$st iteration. Now, if at the end of the $k+1$st iteration there is a z such that $C_z \not\subseteq C_z$, we have two cases. Case a, $z = x$. In this case, either `evaluate_node` or `evaluate_fanin` returned an incorrect result, or some cube C_y was merged into C_x and $C_y \not\subseteq C_y$. The correctness of `evaluate_node` and `evaluate_fanin` has been established, and $C_y \subseteq C_y$ by the invariance assumption, so the proof holds if $z = x$. If $z \neq x$, we have two cases. Either C_z was changed by the action of `propagate_dataflow`, and the item incorrectly added to C_z was in C_x But we must have then that $C_z \in D_x$, i.e., that $x \in C_z$. Since $C_x \subseteq C_x$ through the $k+1$st iteration from above, and since C_z is correct and contains x through the kth iteration, we must have that $C_x \subseteq C_z$. In the second case, $z = \overline{y}$, where y is the literal popped off the stack with C_x, and we know from (C.11) and (C.10) that $C_x \supseteq C_y \Rightarrow C_{\overline{y}} \supseteq C_{\overline{x}}$, and so done. [1] ■

C.6 Complexity Analysis

There are two separate analyses: zero propagation and evaluation. Each literal can have its dataflow cube zeroed at most once, and so the cost of

[1] A careful reader will note that this argument is not quite valid, since both C_x and C_z can change during the $k+1$st iteration in the backward evaluation loop. However, it is easy to extend this argument by arguing that one of the C variables must be the *first* to violate the containment condition; these break down to the two cases above, and are dispatched in the fashion shown; the invalid argument given above has been retained, since it is somewhat clearer than the similar, valid argument.

zeroing the cubes is $O(n)$, where n is the number of cubes in the network, plus the cost of traversing the transitive fanout of each node in the the dataflow graph, which is bounded above by the number of edges of this graph. This in turn is the total number of implications discovered, m, which is bounded above by $O(n^2)$ but should in the usual case be $O(n)$. Hence we argue that this is $O(n+m)$.

If we denote the maximum number of fanins of the nodes in the network as f_1, and the maximum number of fanouts as f_2, and the maximum number of cubes as c, and the maximum number of implicants in any dataflow cube (the maximum size of any $|C_x|$) as d_1, then we see that `evaluate_node(x, y)` and `evaluate_fanin(y, x)` are both $O(cf_1+d_1)$. Clearly f_1, f_2, and d_1 are all $\leq n$, but in general this provides a very loose bound. Similarly, if we denote the maximum size of any D_x as d_2, this is bounded above by n but is in general small. Dataflow propagation is $O(d_1 d_2)$. Simple cube merging is $O(d_1)$. Since we have a new implication for each iteration through the loop, we have at most $O(m)$ iterations. There are $O(f_2)$ evaluations in each loop, and one propagation. Hence the total cost of finding m implications is bounded above by $O((cf_1 f_2 + f_2 d_1 + d_2 d_1)m)$. Note that each of these quantities should be in general small, and so we expect the average-case running time to be linear in n, though the worst-case running time, obtained in the case of a flat network (a network where every node is a primary input or a primary output), is $O(n^4)$. Of course, for such a network, one would hardly wish for dataflow analysis.

C.7 Efficiency

A loose lower bound for the problem is clearly $\Omega(m)$, which is considerably less than the $O((cf_1 f_2 + f_2 d_1 + d_2 d_1)m)$ upper bound derived in the previous section. Both are loose to some degree, but that the Θ bound is likely to be closer to the upper than the lower bound. However, further research is required to derive a tighter lower bound for this problem.

It is easy to see that the algorithms derived here are more efficient than a naive implementation suggested by the recurrence relations:

```
while(changing)
    changing = 0;
    foreach node x
        C'_x = C_x;
```

```
        foreach y ∈ C_x
            C'_x = C'_x ∪ C_y;
            foreach fanout z of y
                if z_{C_x} = 1 C'_x = C'_x ∪ C_z;
                if z_{C_x} = 0 C'_x = C'_x ∪ C_{z̄};
            if y_{C_x} = zf C'_x = C'_x ∪ C_z;
        if(C'_x ≠ C_x) changing = 1;
    foreach node x
        C_x = C'_x;
```

There are potentially $O(m)$ iterations through the outer loop, since one can have as many as one iteration per implication. There are clearly $O(n)$ iterations of the second loop. There are $O(d_1)$ iterations of the third loop. The union operation is also $O(d_1)$. There are $O(f_2)$ fanouts of y, and each evaluation is $O(cf_1 + d_1)$. This implementation is clearly $O(mnd_1 f_2(cf_1 + d_1))$, or $O(n^6)$ in the worst case.

C.8 Sparse Matrix Implementation

When these algorithms were implemented in MisII, it was found that the implementation was excessively slow. Given a CPU time limit of one hour on a Vax 8800, only the circuit C17 of the ISCAS benchmarks completed, in roughly 10 CPU minutes. Profiling revealed that the vast majority of time was spent in the low-level cofactor routines, which ideally should be $O(cf_1 + f_1 + d_1)$, but which may be $O(cf_1 d_1)$ in an implementation not designed with this application in mind. As a result, a sparse matrix implementation was done. In this implementation, an index is assigned to each literal. The row of the matrix corresponding to x represents C_x and the column corresponding to x represents D_x. The cofactoring process is simulated in a straightforward manner. Though the complexity is left unchanged by this implementation, the actual running time of the algorithms are substantially reduced.

C.9 An Improvement

A consistency equation may be considered. If $x \Rightarrow z$, $y \Rightarrow \overline{z}$, it must follow that $x \Rightarrow \overline{y}$ and $y \Rightarrow \overline{x}$, for if x and y, then z and $\overline{z}$, absurd. This gives the equation:

$$x \in C_y^n \quad \text{if} \quad \exists z \in C_x^{n-1} \text{ and } \overline{z} \in C_y^{n-1} \tag{C.15}$$

This equation can actually be deduced from (C.10) and (C.11), and as a result is embedded in the procedure discussed so far. However, there is an interesting corollary. When $C_x = 0$, then $x = 1$ is impossible: x is stuck at 0. Hence *every* node may imply $x = 0$. We write:

$$x \in C_y^n \quad \text{if} \quad C_{\overline{x}}^{n-1} = 0 \tag{C.16}$$

This may be incorporated by modifying `propagate_zero` as follows:

```
propagate_zero(x)
{
    stack = tfo_collect(x);
    while((z = pop_stack(stack)) != nil) {
        C_z = 0;
        foreach literal y ≠ z
            C_y = C_y ∪ C_z̄
    }
}
```

Note this does not affect the complexity of the algorithm.

C.10 Results

The algorithms were implemented and tested on the well-known ISCAS benchmarks. The number of implications discovered is given, as well as the number of dataflow cubes that went to 0. This latter number is of very great interest, for the literals corresponding to these cubes cannot occur and hence the corresponding variable can be set to the opposite value in the circuit (for example, if $C_y = 0$, then every occurrence of y can be replaced by the constant 0; if $C_{\overline{y}} = 0$, then y can be replaced by the constant 1. If both C_y and $C_{\overline{y}} = 0$, the circuit is trivial. Run times are given in seconds on a Vax 8650.

Results and Times for the ISCAS Benchmarks

Circuit	Implicants	Zeroed Dataflow Cubes	Seconds
C1355	26400	0	445.1
C17	25	0	0.4
C1908	35951	0	354.9
C2670	47514	3	614.0
C3540			
C432	1826	0	18.8
C499	6040	0	56.4
C5315	70750	1	535.2
C6288	16070	17	158.9
C7552			
C880	5071	0	32.9

C.11 Extensions

C.11.1 Extending Arbitrary Cubes

The relations (C.7)-(C.11) may apply to any sets of asserted literals; the sets which begin with the initial assertion of a single literal is an interesting seed subcase. In general, however, we are interested in the behaviour of a circuit in response to a *set* C of asserted literals, not merely a single literal, and there is no reason to restrict C to satisfy the condition $C \subseteq C_x$ for some x. The problem is this: given some initial cube C^0 we wish to find some maximal cube of implications from the set of assertions contained in C^0.

Now, clearly one method is to take the union of the data flow sets implied by the cube:

$$C = \bigcup_{x \in C^0} C_x \tag{C.17}$$

with of course the caveat:

$$C = 0 \text{ if } \exists x \in C^0 \text{ such that } C_x = 0 \tag{C.18}$$

where C_x is, as before, the fixed point of relations (C.7)-(C.11). However, we can in general do somewhat better if $C^0 \not\subseteq C_x$ for some x. The union relation merely reflects the dataflow propagation of the C_x sets into the new cube. However, other implications may be derived from analogues to (C.7)-(C.9). From this we derive the equations:

$$
\begin{aligned}
y \in C^n &\quad \text{if} \quad y_{C^{n-1}} = 1 && \text{(C.19)}\\
y \in C^n &\quad \text{if} \quad \exists t \in C^{n-1} \text{ and } t_{C^{n-1}} = yf \text{ some } f && \text{(C.20)}\\
y \in C^n &\quad \text{if} \quad \exists \bar{t} \in C^{n-1} \text{ and } t_{C^{n-1}} = \bar{y} + f \text{ some } f && \text{(C.21)}\\
y \in C^n &\quad \text{if} \quad \exists t \in C^{n-1} \text{ and } y \in C_t && \text{(C.22)}
\end{aligned}
$$

which are fairly clearly analogues of equations (C.7)-(C.10). An analogue to equation (C.11) is not required, since the contrapositive is only closed under the action of individual literals and hence is closed by the existing sets C_x.

Theorem C.11.1 *Consider any C^0. Let C be obtained as a fixed point of relations (C.19)-(C.22). Then if $y(\bar{y}) \in C$, y is set to $1(0)$ when the cube C^0 is set.*

Proof: Attach a node $\eta = C^0$ to the network, where η does not fan out. Observe that the relations (C.19)-(C.22) are identical to those obtained by relations (C.7)-(C.11) for such a new node, and apply the results of Lemma C.3.1. ■

The code to implement this is quite straightforward. One merely adds an extra row to the sparse matrix developed above, and implements a modified version of the main loop developed above; this modified version simply removes extraneous code relating to dataflow propagation of the cube C, whose dataflow implications clearly do not fan out.

```
while((C, y) = pop_stack(evaluation_stack)) {
    if C_x = 0 return 0;
    foreach fanout w of y {
        if((w has a literal w1 ∈ C) { (w1 is either w or w̄)
           new_cube = evaluate_fanin(w1, x);
           foreach literal z in new_cube
               merge C‡ into C;
        }
        phase = evaluate_node(w, x);
        if(phase != 2){
           w1 is the literal suggested by w and phase;
           merge C_w1 into C;
        }
```

```
    }
    new_cube = evaluate_fanin(y, C);
    foreach literal z in new_cube
        merge C‡ into C;
}
```

C.11.2 The Fanout Care Set and the Test Function

For some applications, (e.g., testing), one not only wishes the circumstances under which some literal may be set, but also the circumstances under which the setting of that variable may propagate to the output. This function, known variously as the *fanout care* condition or the *boolean difference*, may be written, for each variable x:

$$\sum_i f^i_x \oplus f^i_{\bar{x}} \tag{C.23}$$

and the implications which this required may be found by assigning the node η of the previous section to this value and performing the implications in the manner above.

Appendix D

Precharged, Unate Circuits

CMOS circuitry comes in two basic flavours: dynamic and static. Static circuitry, also called *restoring* logic, is fully complementary. Each gate consists of a *pulldown* network of NMOS transistors connecting the gate output to ground, and a *pullup* network of PMOS transistors connecting the output to power, or, as it is better known, V_{dd}. The pullup network is the dual of the pulldown network; hence exactly one of the two networks is conducting at any time, and so the output is affirmatively driven to its logic value. Hence a static gate is responsive to values on its inputs at all times; in logical terms, the gate may be thought of as an ideal logic switch.

Static CMOS gates obviously involve some redundancy at the transistor level, since either network is sufficient to compute the logic function. In the very early days of CMOS design (the late 1970's and early 1980's), a fairly dubious line of reasoning held that this wasted area, due to the large separation required between the p- and n-regions in CMOS technology[1].

The immediate idea was to eliminate one of the two transistor networks; the problem, therefore, was how to drive the logic to the appropriate value when the missing transistor network was putatively active. In 1982, Krambeck, Lee and Law [54] proposed to use the capacitive properties of the gate terminal of the MOS switch and of interconnect to realize its function; i.e., the fact that the wires and gate terminals of MOS transistors act as storage elements when disconnected from either

[1]this separation is required to avoid an electrical problem known as *latchup*; this problem is not germane to the subject of this thesis–for details, see [90]

power or ground.

Consider the case where the pullup network is eliminated. In this case, the gate cannot be driven to a logic 1 when required, though it can still be driven to a logic 0. Krambeck and his co-workers realized that in the case where a static gate would be driven to 1, the gate missing the pullup network would be disconnected from either power or ground and so would retain the value left on it. Hence if this value was 1, the gate would compute its correct value in this case. In the case where the pulldown network was active, the gate would be set to 0 as usual.

This design leads to the picture of a two-phase design. During the first or *precharging* phase, the pulldown network is disabled and the output of the gate is connected to power, so the storage element inherent in MOS logic is set to logic 1. During the second or *evaluation* phase, the network is enabled and the connection to power disabled.

The enabling and disabling of the relevant connections is fairly easy, as is shown in figure D.1. In this figure, a single PMOS transistor, controlled by φ, runs between power and the output. Similarly, a single NMOS transistor, also controlled by φ, runs between ground and the pulldown network.

Now, when φ is low, the PMOS transistor is conducting and the NMOS transistor is not; hence the output is connected to power and the pulldown network is disconnected from ground, i.e., is disabled – φ is low on the precharge phase. When φ is high, the PMOS transistor is not conducting and the output is isolated from power, and the pulldown network is connected to ground – i.e., the pulldown network is enabled.

Note that there is only a finite amount of charge stored on the output of any gate; hence, it is critical that the pulldown network only be conducting during the phase if its final state is conducting; otherwise, during the period it is conducting, the output can be driven to ground; this is a fatal error if the final value is 1, as it is if the pulldown network is nonconducting.

This has three immediate consequences. First, the initial state of the pulldown network must be non-conducting – all the transistors that may change during evaluation must be initially off. Further, no transistor may turn on during evaluation unless its final value is on; i.e., the gate must be hazard-free to function correctly. Finally, since all transistors that may change must be precharged to their off state, and since the primary inputs may not be precharged, the evaluation phase may not begin until the primary inputs have all reached their stable value.

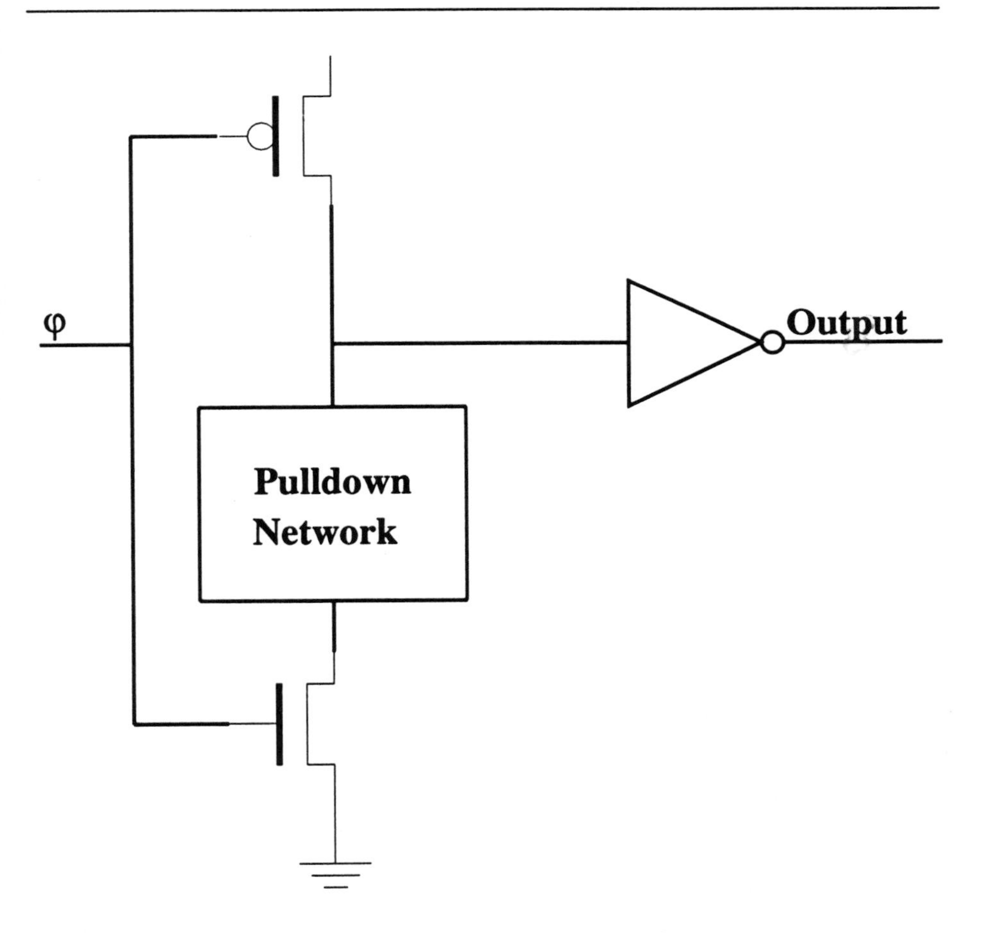

Figure D.1: Generic Dynamic Gate

The hazard-free property is assured by construction. For the other, note that NMOS transistors are conducting for logic 1, nonconducting for logic 0. Hence the inputs to a gate must be precharged to 0. However, the precharge state of this gate is 1; if the inputs are gates like it, this will not do.

In Krambeck et. al.'s work, this is handled by attaching a static inverter to each gate; the gate itself fans out only to the static inverter, which is the real output of the gate. The gate is therefore precharged to 0 (since the precharge state of the inverter is obviously the inverse of that of the gate proper), and the gate can only undergo $0 \rightarrow 1$ transitions during evaluation. If this is done, this logic is called DOMINO. Now, since the basic MOS gate is inverting, the basic DOMINO gate is non-inverting; for the pulldown network will in general realize the complement of some (small) AND/OR network; the static inverter forces the DOMINO gate to realize the complement of the pulldown network, i.e. the AND/OR network. Note that adjustments need be made to the logic to realize an arbitrary function, since only the primary inputs of the circuit may appear in inverted form. However, it is easy to transform an arbitrary network into a network consisting of only AND and OR gates, where the primary inputs appear in both inverted and noninverted form.

We illustrate the basic properties of DOMINO logic with a simple example, shown in figure D.2. This gate realizes the function ab when φ is high.

It eventually developed that DOMINO logic saved little area, at least when laid out with an automated synthesis tool [41]. However the same work also showed that DOMINO logic switched about 30% faster than static CMOS.

The lack of inverting devices was a difficulty with DOMINO logic, since the technique used to transform an arbitrary network into a network of only AND and OR gates could potentially double the gate count, resulting in a circuit unacceptably large. Another approach, was devised by Goncalves and De Man and detailed in [33] and [29]. In their circuit style, called NORA, *only* inverting gates were permitted. NORA circuits consist of alternating *n*- and *p-type* gates. An *n-type* gate is simply a DOMINO gate *sans* the static inverter. A *p-type* gate is a gate consisting of a pullup rather than a pulldown network. The precharge transistor on a p-type gate is an NMOS transistor; the evaluate transistor is a PMOS transistor. Since the PMOS transistor is active when its control is at logic 0, if the precharge and evaluate transistors on the n-

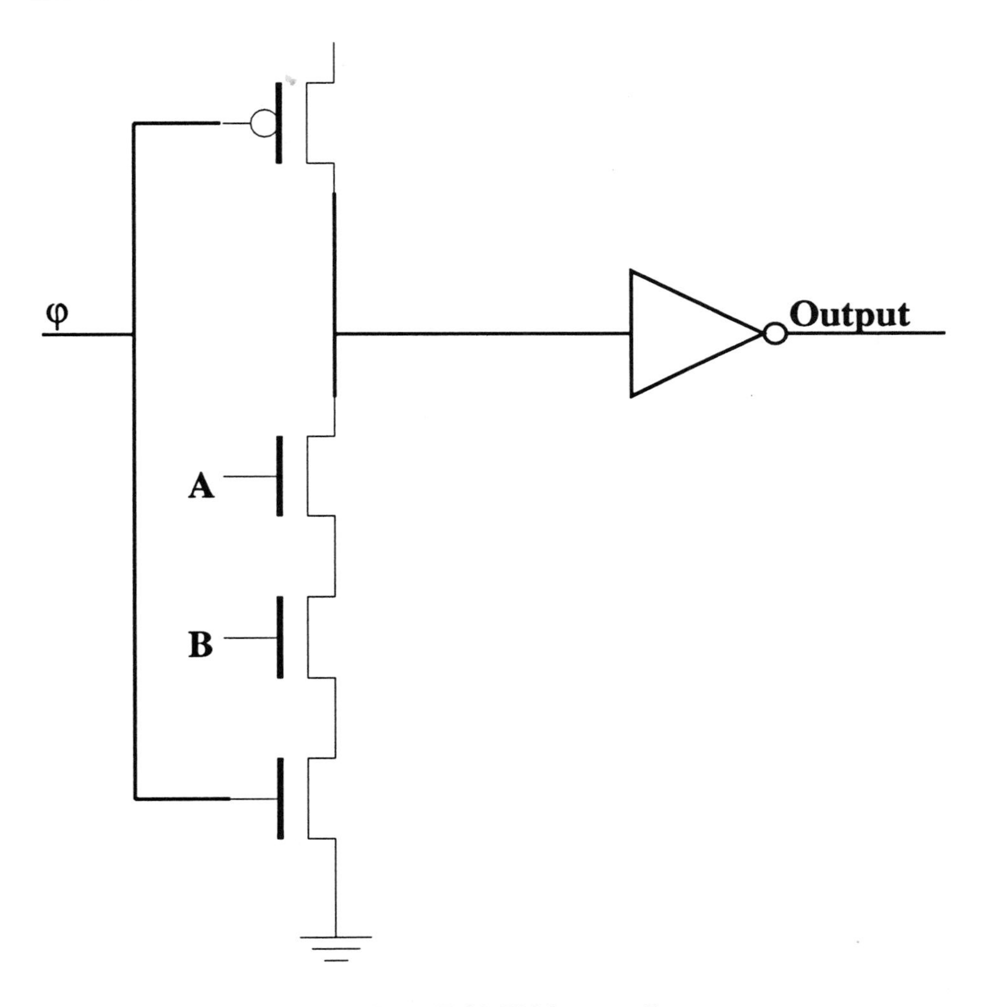

Figure D.2: DOMINO AND Gate

type gate are controlled by φ, the corresponding transistors on the p-type gate is controlled by $\overline{\varphi}$. Static inverters are classed as p-type gates if they are fed by an n-type gate, and n-type gates if they are fed by a p-type gate. Hence one may say that the output of a p-type gate is precharged to logic 0, and the output of an n-type gate is precharged to logic 1. Further, the same rule that input transistors must be precharged to their off state applies to NORA logic; the off-state for p-type gate inputs is logic 1, the off-state for n-type gate inputs is logic 0; this argument leads to the conclusion that n-type gate outputs can feed only p-type gates, and vice-versa.

Note that any NORA circuit can be transformed into an equivalent DOMINO circuit by adding inverters to the output of each gate and moving the pullup trees in p-type gates to the pulldown n-type well.

An example of a generic NORA network is shown in figure D.3.

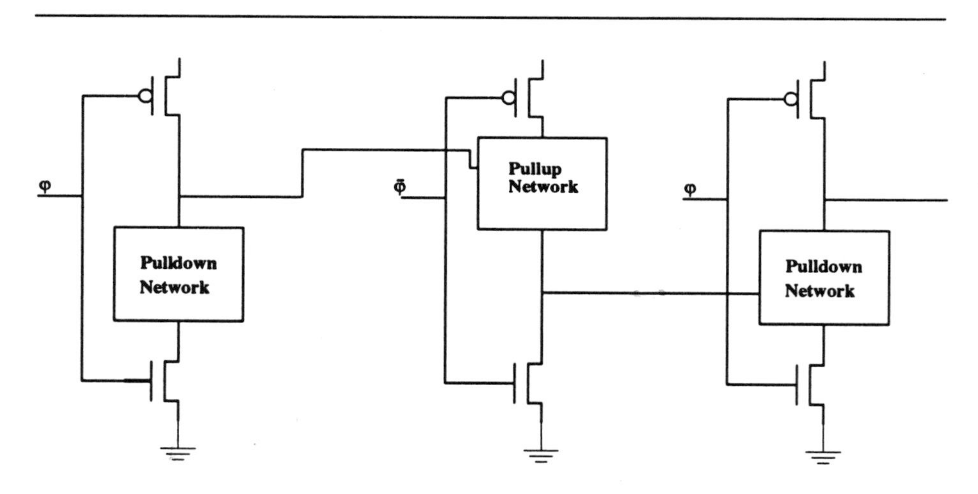

Figure D.3: Generic NORA Gate

A disadvantage of NORA circuits is that p-type transistors are typically slower to switch than n-type, due to decreased electron mobility in the p-well. Further, the fact that p-type gates can only feed n-type gates and vice-versa makes NORA circuits very hard to design.

A final type of precharged circuit was introduced by Heller et al [39].

In this form of logic, called *Differential Cascode Voltage Switch* or, more simply, DCVS, the pulldown trees for both a function and its complement are implemented; a DCVS gate is simply two merged DOMINO gates back-to-back. This form of logic has the advantage that both complemented and uncomplemented logic is available, without NORA's slow p-type gates and difficulty. However, naive implementations of DCVS require as many transistors as full static; further, both each signal and its complement must be routed throughout the network in every DCVS implementation. These two factors indicate that the resulting circuit must be approximately twice the area of a static implementation; however, the flexibility of the DCVS design style and clever physical design have combined to offset this penalty[93].

In general, one can do somewhat better than naive implementations of DCVS. It was realized that DCVS circuits mapped nicely onto Bryant's graph-based representations of Boolean functions, called *Boolean Decision Diagrams* or *BDDs*. A BDD is a rooted, binary, directed acyclic graph with two leaves, where each non-leaf node is labelled with the name of a variable, and each edge is labelled either 1 or 0. The leaf nodes are labelled 1 and 0.

A BDD is intended to model the computation of a boolean function, given assignments of the variables, as a traversal of the graph. At each node, the edge labelled i is followed when the variable that labels that node is set to i. The value of the function is the label on the leaf at which the variable terminates.

Consider the edge labelled 1 originating in node labelled x to be a transistor that is active when $x = 1$; this clearly corresponds to an NMOS transistor controlled by x. Similarly, the edge labelled 0 corresponds to an NMOS transistor controlled by $\bar{x}$. In this case, there is a path from the node labelled 0 to the root iff f evaluates to 0, and a path from 1 to the root iff f evaluates to 1, i.e., $\bar{f}$ evaluates to 0. If the root is then directly connected to ground, it is clear that the function is correctly implemented (that is, there is a connecting path from ground to the appropriate node) if the node labelled 1 is replaced by a node labelled $\bar{f}$, and the node labelled 0 is replaced by a node labelled f. The derivation of this connection tree is shown in figure D.4. The full DCVS implementation, with the appropriate precharging logic, is shown in figure D.5. Note that, as with DOMINO implementations, the DCVS nodes must all be precharged low.

Note that each form of logic shown here has the property that the

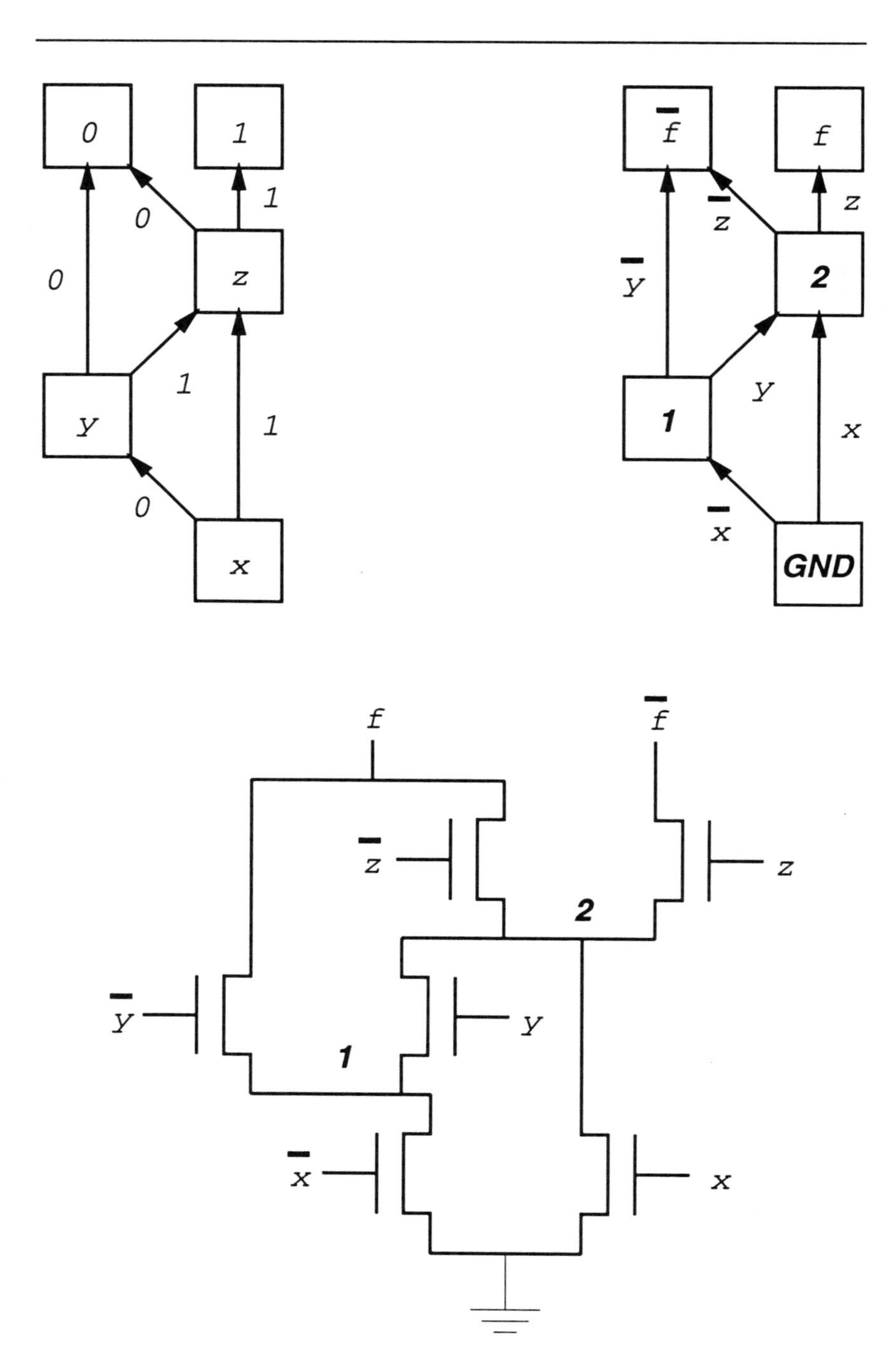

Figure D.4: BDD and DCVS Representation of $f = xz + \bar{x}yz$

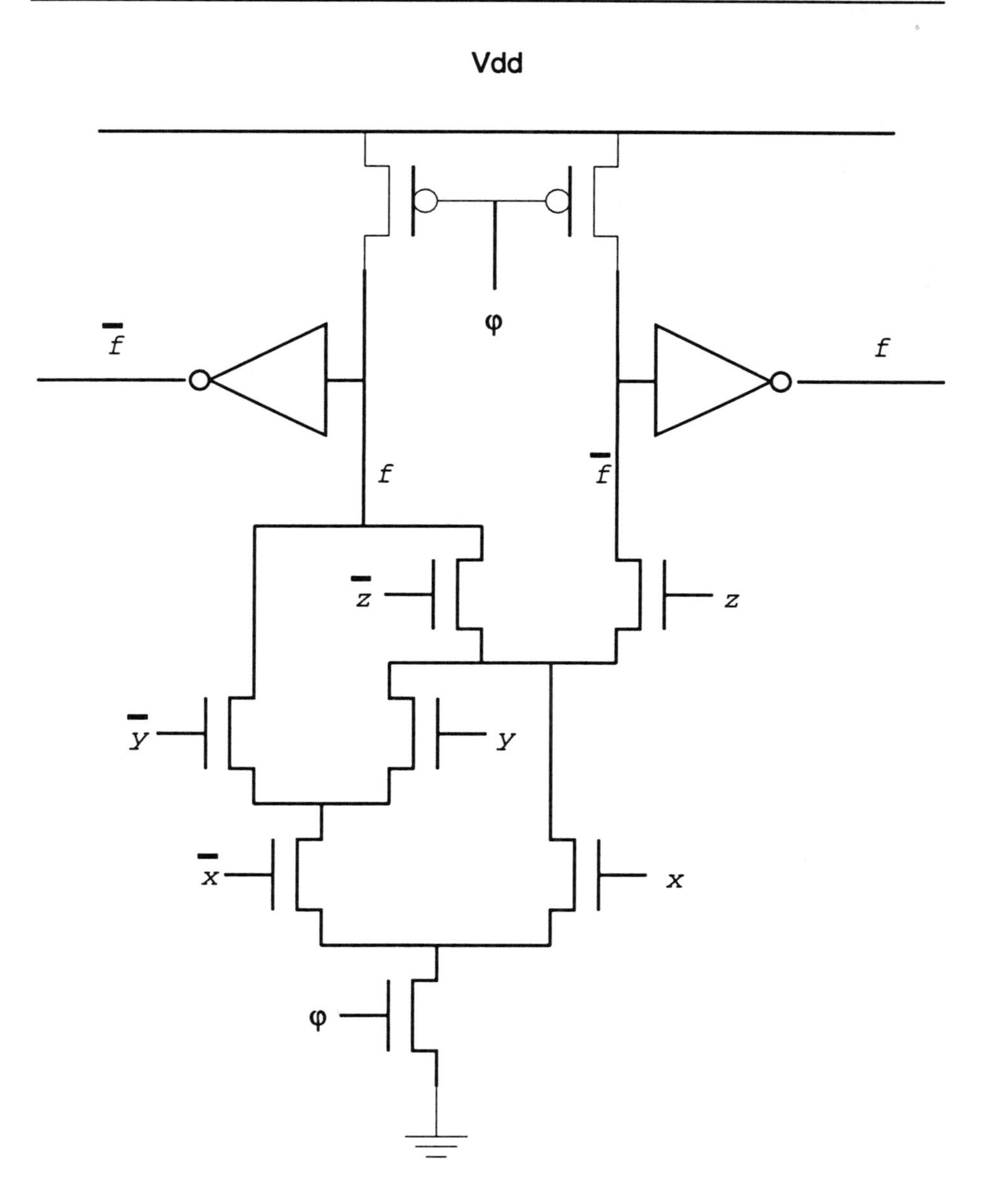

Figure D.5: Full DCVS Implementation $f = xz + \bar{x}yz$

resulting gate can only change from 1 to 0 or from 0 to 1. Such gates are called *unate.* From the discussion above, it is clear that this is an inherent property of dynamic logic.

Interestingly, though the DCVS gate is unate it may realize any boolean function; e.g, the function $xz + \bar{x}yz$ is easily realizable by a DCVS gate. The reason that this is possible is that the literals x and $\bar{x}$ are carried on independent wires and hence are indistinguishable at the gate level from independent variables. The central attraction of DCVS technology is that it combines the strong timing properties of DOMINO demonstrated in this thesis and the full power of the static boolean gate.

Bibliography

[1] S. B. Akers. On a Theory of Boolean Functions. *J. SIAM*, 1959.

[2] K. A. Bartlett, D. G. Bostick, G. D. Hachtel, R. M. Jacoby, M. R. Lightner, P. H. Moceyunas, C. R. Morrison, and D. Ravenscroft. BOLD: A multi-level logic optimization system. In *IEEE International Conference on Computer-Aided Design*, 1987.

[3] K. A. Bartlett, R. K. Brayton, G. D. Hachtel, R. M. Jacoby, R. Rudell, A. Sangiovanni-Vincentelli, and A. Wang. Multi-level logic minimization using implicit don't cares. *IEEE Transactions on CAD*, 1988.

[4] Romy L. Bauer, Jiayuan Fang, Antony P-C Ng, and Robert K. Brayton. XPSim: A MOS VLSI circuit simulator. In *IEEE International Conference on Computer-Aided Design*, 1988.

[5] Romy L. Bauer, Antony P-C Ng, Arvind Raghunathan, Mark W. Saake, and Clark D. Thompson. Simulating MOS VLSI circuits using SuperCrystal. In *VLSI '87*, 1987.

[6] J. Benkoski, E. Vanden Meesch, L. Claesen, and H. DeMan. Efficient algorithms for solving the false path problem in timing verification. In *IEEE International Conference on Computer-Aided Design*, 1987.

[7] C. L. Berman, J. L. Carter, and K. F. Day. The fanout problem: From theory to practice. In *Decennial CalTech VLSI Conference*, 1989.

[8] L. Berman, L. Trevillyan, and W. Joyner. Global flow analysis in automated logic design. *IEEE Transactions on Computers*, January 1986.

[9] Daniel Brand. Redundancy and don't cares in logic synthesis. *IEEE Transactions on Computers*, October 1983.

[10] Daniel Brand. Personal communication, 1988.

[11] Daniel Brand and Vijay S. Iyengar. Timing analysis using functional analysis. Technical Report RC 11768, IBM Thomas J. Watson Research Center, Yorktown Heights, New York, 10598, 1986.

[12] Daniel Brand and Vijay S. Iyengar. Timing analysis using functional analysis. In *IEEE International Conference on Computer-Aided Design*, 1986.

[13] R. K. Brayton, E. Detjens, S. Krishna, T. Ma, P. McGeer, L. Pei, N. Phillips, R. Rudell, R. Segal, A. Wang R. Yung, and A. L. Sangiovanni-Vincentelli. Multiple level logic optimization system. In *IEEE International Conference on Computer-Aided Design*, 1986.

[14] R. K. Brayton, G. Hachtel, C. T. McMullen, and A. L. Sangiovanni-Vincentelli. *Logic Minimization Algorithms for VLSI Synthesis.* Kluwer Academic Publishers, 1984.

[15] R. K. Brayton, R. L. Rudell, A. L. Sangiovanni-Vincentelli, and A. Wang. MIS: A multi-level logic synthesis system. *IEEE Transactions on CAD*, 1987.

[16] R. K. Brayton, F. Somenzi, and E. M. Sentovich. Don't cares and global flow analysis of boolean circuits. In *IEEE International Conference on Computer-Aided Design*, 1988.

[17] M. A. Breuer. The effects of races, delays, and delay faults on test generation. *IEEE Transactions on Computers*, 1974.

[18] M. A. Breuer and R. Lloyd Harrison. Procedures for eliminating static and dynamic hazards in test generation. *IEEE Transactions on Computers*, October 1974.

[19] R. E. Bryant. Graph-based algorithms for Boolean function manipulation. *IEEE Transactions on Computers*, 1981.

[20] R. E. Bryant. MOSsim: A switch-level simulator for MOS LSI. In *Design Automation Conference*, 1981.

[21] H. C. Chen and D. H. C. Du. On the critical path problem. In *ACM International Workshop on Timing Issues in the Specification and Synthesis of Digital Systems (τ '90)*, 1990.

[22] James J. Cherry. PEARL: A CMOS timing analyzer. In *Design Automation Conference*, 1988.

[23] S. Cook. On the complexity of theorem-proving procedures. In *ACM Symposium on the Theory of Computing*, 1971.

[24] Ewald Detjens, Gary Gannot, R. L. Rudell, A. L. Sangiovanni-Vincentelli, and A. Wang. Technology mapping in mis. In *IEEE International Conference on Computer-Aided Design*, 1987.

[25] J. T. Deutsch and A. R. Newton. A multiprocessor implementation of relaxation-based electrical circuit simulation. In *Design Automation Conference*, 1984.

[26] David H. C. Du, Steve H. C. Yen, and S. Ghanta. On the general false path problem in timing analysis. In *Design Automation Conference*, 1989.

[27] E. B. Eichelberger. Hazard detection in combinational and sequential switching circuits. *IBM Journal of Research and Development*, March 1965.

[28] Jiayuan Fang. The approximate exponential function method for circuit simulation. Technical report, Electronics Research Laboratory, UC-Berkeley, 1987.

[29] V. Friedman and S. Liu. Dynamic logic CMOS circuits. *IEEE Journal of Solid State Circuits*, 1984.

[30] Michael R. Garey and David S. Johnson. *Computers and Intractability: A Guide to the Theory of NP-Completeness.* W. H. Freeman and Company, 1979.

[31] C. Thomas Glover and M. Ray Mercer. A method of delay fault test generation. In *Design Automation Conference*, 1988.

[32] Prabhakar Goel. An implicit enumeration algorithm to generate tests for combinational logic circuits. *IEEE Transactions on Computers*, 1980.

[33] Nelson F. Goncalves and Hugo J. DeMan. NORA:a racefree dynamic CMOS technique for pipelined logic structures. *IEEE Journal of Solid State Circuits*, 1983.

[34] David Gries. *The Science of Programming*. Springer-Verlag, 1981.

[35] G. Hachtel, R. Jacoby, K. Keutzer, , and C. Morrison. On the relationship between area optimization and multifault testability of multilevel logic. In *International Workshop on Logic Synthesis*, 1989.

[36] G. Hachtel, R. Jacoby, and P. Moceyunas. On computing and approximating the observability don't-care set. In *International Workshop on Logic Synthesis*, 1989.

[37] G. Hachtel, R. Jacoby, P. Moceyunas, and C. Morrison. Performance enhancements in BOLD using "implications". In *IEEE International Conference on Computer-Aided Design*, 1988.

[38] D. Hathaway, L. H. Trevillyan, C. L. Berman, and A. S. LaPaugh. Efficient techniques for timing correction. In *IEEE International Symposium on Circuits and Systems*, 1990.

[39] L. G. Heller, W. R. Griffin, J. W. Davis, and N. G. Thoma. Cascode voltage switch logic: A differential CMOS logic family. In *IEEE International Solid State Circuits Conference*, 1984.

[40] Robert B. Hitchcock. Timing verification and the timing analysis program. In *Design Automation Conference*, 1982.

[41] Mark Hofmann. *Automated Synthesis of Multi-Level Logic in CMOS Technology*. PhD thesis, Department of Electrical Engineering and Computer Science, University of California at Berkeley, 1982.

[42] V. M. Hrapcenko. Depth and delay in a network. *Soviet Math. Dokl.*, 1978.

[43] N. Jouppi. TV: An nMOS timing analyzer. In *Third Caltech VLSI Conference*, 1983.

[44] N. Jouppi. Deriving signal flow direction in MOS VLSI. *IEEE Transactions on CAD*, july 1987.

[45] N. Jouppi. Timing analysis and performance improvement of MOS VLSI designs. *IEEE Transactions on CAD*, may 1987.

[46] R. M. Karp. Reducibility among combinatorial problems. In *ACM Symposium on the Theory of Computing*, 1971.

[47] K. Keutzer and M. Vancura. Timing optimization in a logic synthesis system. In *International Workshop on Logic and Architectural Synthesis in Silicon Compilers*, 1988.

[48] Kurt Keutzer, Sharad Malik, and Alexander Saldanha. Is redundancy necessary to reduce delay? In *Design Automation Conference*, 1990.

[49] Y. H. Kim. *Accurate Timing Verification for VLSI Designs.* PhD thesis, Department of Electrical Engineering and Computer Science, University of California at Berkeley, 1989.

[50] Y. H. Kim, S. H. Hwang, and A. R. Newton. Electrical-logic simulation and its application. *IEEE Transactions on CAD*, January 1989.

[51] T. W. Kirkpatrick and N. Clark. PERT as an aid to logic design. *IBM Journal of Research and Development*, 1966.

[52] T. W. Kirkpatrick and N. Clark. PERT as an aid to logic design. *IBM Journal of Research and Development*, 1966.

[53] D. Knuth. *The Art of Computer Programming.* Addison-Wesley, 1973.

[54] R. H. Krambeck, C. M. Lee, and H-F. S. Law. High-speed compact circuits with CMOS. *IEEE Journal of Solid State Circuits*, 1982.

[55] T. Larrabee. Efficient generation of test patterns using Boolean difference. In *International Test Conference*, 1989.

[56] Jean Davies Lesser and John J. Shedletsky. An experimental delay fault test generator for LSI logic. *IEEE Transactions on Computers*, 1980.

[57] Chin Jen Lin and Sudhakar M. Reddy. On delay fault testing in logic circuits. *IEEE Transactions on CAD*, 1987.

[58] S. Malik, A. Wang, R. K. Brayton, and A. L. Sangiovanni-Vincentelli. Logic verification using binary decision diagrams in a logic synthesis environment. In *IEEE International Conference on Computer-Aided Design*, 1988.

[59] Hugo De Man. Personal communication, 1988.

[60] P. C. McGeer. *On the Interaction of Functional and Timing Behaviour of Combinational Logic Circuits.* PhD thesis, Department of Electrical Engineering and Computer Science, University of California at Berkeley, 1989.

[61] P. C. McGeer, A. Saldanha, P. Stephan, R. K. Brayton, and A. L. Sangiovanni-Vincentelli. Timing analysis and delay-fault test generation using path recursive functions. In *ACM International Workshop on Logic Synthesis*, 1991.

[62] Patrick C. McGeer and Robert K. Brayton. Efficient algorithms for computing the longest viable path in a combinational network. In *Design Automation Conference*, 1989.

[63] Patrick C. McGeer and Robert K. Brayton. Provably correct critical paths. In *Decennial CalTech VLSI Conference*, 1989.

[64] Patrick C. McGeer and Robert K. Brayton. Hazard prevention in combinational circuits. In *Hawaii International Conference on the System Sciences*, 1990.

[65] Patrick C. McGeer and Robert K. Brayton. Timing analysis on precharged-unate networks. In *Design Automation Conference*, 1990.

[66] Patrick C. McGeer, Robert K. Brayton, Richard L. Rudell, and Alberto L. Sangiovanni-Vincentelli. Extended stuck-fault testability for combinational networks. In *MIT Conference on Advanced Research in VLSI*, 1990.

[67] Patrick C. McGeer, Robert K. Brayton, and Alberto L. Sangiovanni-Vincentelli. Performance enhancement through the generalized bypass transform. In *International Workshop on Logic Synthesis*, 1991.

[68] T. M. McWilliams. Verifiction of timing constraints on large digital systems. In *Design Automation Conference*, 1980.

[69] L. W. Nagel. SPICE2: A computer program to simulate semiconductor circuits. Technical Report UCB/ERL M75/520, Electronics Research Lab, University of California at Berkeley, 1975.

[70] A. R. Newton and A. L. Sangiovanni-Vincentelli. Relaxation-based electrical simulation. *IEEE Transactions on Electronic Devices*, 1983.

[71] John K. Ousterhout. Crystal: A Timing Analyzer for nMOS VLSI Circuits. In *Third Caltech VLSI Conference*, 1983.

[72] John K. Ousterhout. Switch-level delay models for digital MOS VLSI. In *Design Automation Conference*, 1984.

[73] John K. Ousterhout. A switch-level timing verifier for digitial MOS VLSI. *IEEE Transactions on CAD*, July 1985.

[74] P. Penfield, Jr. and J. Rubinstein. Signal delay in RC tree networks. In *Design Automation Conference*, 1981.

[75] S. Perremans, L. Claesen, and H. De Man. Static timing analysis of dynamically sensitizable paths. In *Design Automation Conference*, 1989.

[76] T. Quarles. *The* SPICE3 *Circuit Simulator.* PhD thesis, Department of Electrical Engineering and Computer Science, University of California at Berkeley, 1989.

[77] J. P. Roth. Diagnosis of automata failures: A calculus and a method. *IBM J. Res. Develop*, 1966.

[78] J. Rubinstein, P. Penfield, Jr., and M. A. Horowitz. Signal delay in RC tree networks. *IEEE Transactions on CAD*, July 1983.

[79] Robert Sedgewick. *Algorithms.* Addison-Wesley, 1983.

[80] Frederick F. Sellers, Jr., M. Y. Hsiao, and L. W. Bearnson. Analyzing errors with the Boolean difference. *IEEE Transactions on Computers*, 1968.

[81] C. E. Shannon. The synthesis of two-terminal switching function. *Bell System Technical Journal*, 1948.

[82] Kanwar Jit Singh, Albert R. Wang, Robert K. Brayton, and Alberto L. Sangiovanni-Vincentelli. Timing optimization of combinational logic. In *IEEE International Conference on Computer-Aided Design*, 1988.

[83] G. L. Smith. Model for delay faults based upon paths. In *International Test Conference*, 1985.

[84] Herve J. Touati, Cho W. Moon, Robert K. Brayton, and Alberto L. Sangiovanni-Vincentelli. Performance-oriented technology mapping. In *MIT Conference on Advanced Research in VLSI*, 1990.

[85] L. Trevillyan and L. Berman. Improved logic optimization using global flow analysis. In *IEEE International Conference on Computer-Aided Design*, 1988.

[86] D. E. Wallace and C. H. Sequin. Plug-in timing models for an abstract timing verifier. In *Design Automation Conference*, 1986.

[87] D. E. Wallace and C. H. Sequin. ATV: An abstract timing verifier. In *Design Automation Conference*, 1988.

[88] D. M. Webber and A. L. Sangiovanni-Vincentelli. Circuit simulation on the connection machine. In *Design Automation Conference*, 1987.

[89] N. Weiner and A. L. Sangiovanni-Vincentelli. Timing analysis in a logic synthesis environment. In *Design Automation Conference*, 1989.

[90] Neil H. E. Weste and Kamran Eshraghian. *Principles of CMOS VLSI Design: A Systems Perspective.* Addison-Wesley, 1985.

[91] J. White and A. L. Sangiovanni-Vincentelli. *Relaxation-based Circuit Simulation.* Kluwer Academic Publishers, 1985.

[92] Steve H. C. Yen, David H. C. Du, and S. Ghanta. Efficient algorithms for extracting the k most critical paths in timing analysis. In *Design Automation Conference*, 1989.

[93] Ellen J. Yoffa and Peter S. Hauge. ACORN: A local customization approach to DCVS physical design. In *Design Automation Conference*, 1985.

Index